Integrated Urban Water Management

Integrated Urban Water Management

Edited by **Herbert Lotus**

New York

Published by Callisto Reference,
106 Park Avenue, Suite 200,
New York, NY 10016, USA
www.callistoreference.com

Integrated Urban Water Management
Edited by Herbert Lotus

International Standard Book Number: 978-1-63239-760-7 (Hardback)

Contents

Permissions

List of Contributors

Preface

In my initial years as a student, I used to run to the library at every possible instance to grab a book and learn something new. Books were my primary source of knowledge and I would not have come such a long way without all that I learnt from them. Thus, when I was approached to edit this book; I became understandably nostalgic. It was an absolute honor to be considered worthy of guiding the current generation as well as those to come. I put all my knowledge and hard work into making this book most beneficial for its readers.

Water is essential for all life forms. Although it is available in abundance, portable water forms a small percentage of all the water resources available. Water management refers to the practice of managing, developing, distributing and planning optimum use of water resources. As population and pollution are increasing rapidly, judicial management of water is the need of the hour. This book traces the progress of this field and highlights some of its key concepts and applications. It strives to provide a fair idea about water management and develop a better understanding of the latest advances within this field. Scientists and students actively engaged in this field will find this book full of crucial and unexplored concepts. Those in search of information to further their knowledge in this area will be greatly assisted by this book.

I wish to thank my publisher for supporting me at every step. I would also like to thank all the authors who have contributed their researches in this book. I hope this book will be a valuable contribution to the progress of the field.

Editor

Water Distribution in the Perspectives of Stakeholders and Water Users in the Tarim River Catchment, Xinjiang, China

Haiyan Peng[1], Niels Thevs[1], Konrad Ott[2]

[1]Institute of Botany and Landscape Ecology, University of Greifswald, Greifswald, Germany
[2]Department of Philosophy, Christian-Albrecht-University Kiel, Kiel, Germany
Email: Thevs@uni.greifswald.de

Abstract

Many river basins in the arid and semi-arid parts of the world are experiencing water scarcity due to water consumption by agriculture resulting in conflicts between upstream and downstream, conflicts between water users, and degradation of the natural ecosystems. The Tarim Basin, Xinjiang, China, has developed into the world's most important cotton production region with 8.85% of the world's production. Under the extremely arid climate with annual precipitation of below 100 mm, the water consumption due to irrigation resulted in water scarcity and conflicts between water users as well as between upstream and downstream. The Tarim river catchment harbors about half of the world's *Populus euphratica* riparian forests, which are impacted by water shortage. Starting in the 1990s, a unified water management system with a quota system for water distribution has been set up. We introduce this unified water management system and analyze how the water distribution works in practice. Ecologists and forestry officials claim more water for environmental flow, whereas water management officials give priority to agricultural, industrial, and domestic water use. The water quotas for downstream regions are frequently not fulfilled, especially during the non-flood season in spring and early summer posing a risk to water users. Water users with financial and political advantages gain more water security than others. The water quotas are annual quotas. These quotas should be differentiated into seasonal quotas, in order to enhance water security for the downstream section of the Tarim all year round.

Keywords

Water Resource Management, Central Asia, Water Scarcity, Land Use, Cotton, Environmental Flow

1. Introduction

Many river basins in the arid and semi-arid parts of the world are experiencing water scarcity due to water consumption by agriculture resulting in conflicts between upstream and downstream, conflicts between water users, and degradation of the natural ecosystems, such as the Aral Sea Basin [1], Jordan [2], Murray-Darling [3], Yellow River [4], Heihe [5], and Tarim Basin [6] [7]. Given the background of water scarcity, increasing populations, and increasing water demands, water distribution in river basins between upstream and downstream as well as between sectors plays an important role especially for river basins under arid and semi arid climate.

The Tarim Basin, located in the Xinjiang Uyghur Autonomous Region (hereinafter referred to as Xinjiang), China, covers an area of 1.02 million km^2, is home to a population of about 9.5 million people, and has turned into the world's most important cotton production region with a total annual cotton lint production of 2.1 million t, *i.e.* 8.85% of the world production, in 2010 [8] [9]. In 2011, the share of the cotton lint production in Xinjiang of the worldwide production climbed to 11% [9] [10]. The catchment of the Aksu and Tarim is the largest river catchment in the Tarim Basin with a river length of 282 km and 1321 km of the Aksu and Tarim, respectively [6]. Half of the cotton in the Tarim Basin is produced along the Aksu and Tarim River and most of the population of the Tarim Basin live in the Aksu and Tarim river catchment [8] [11]. Due to the arid climate with an annual precipitation of 30 to 70 mm in the cropland area along the Aksu and Tarim [7] [12], the cotton production and other agriculture depend on irrigation. The water consumption by irrigation has resulted in periods of water shortage along the midstream and downstream section of the Tarim River during the past ten years [7] [13].

The Tarim Basin also harbors 54% (352,200 ha) of the world's riparian *Populus euphratica* Oliv. forests [14]. Those forests form a mosaic of riparian forests, wetlands, shrub vegetation, and small stands of herbaceous vegetation [15] and provide habitat for wildlife [14]. The *Populus euphratica* forests are the only forests in the Tarim Basin. Those forests and the wetlands are the most productive ecosystems of the drylands in the Tarim Basin [16] [17]. The largest contiguous areas of the *Populus euphratica* forests with associated wetlands and shrub vegetation are located along the Tarim River in the two nature reserves Tarim Shangyou and Tarim Huyanglin, which stretch along the Tarim River in Xayar County between Aral and Yingbaza and downstream from Yingbaza, respectively [18].

Since 1950, through reclamation of land for irrigated agriculture and rapid population growth, partly driven by resettlement of people from other Chinese provinces to Xinjiang, the water withdrawal from all rivers in the Tarim Basin increased. Until the 1970s, the Aksu, Hotan, and Yarkant River were tributaries, which permanently discharged into the Tarim. From the 1970s onward, the latter two rivers have become disconnected from the Tarim. Also at the beginning of the 1970s, the 320 km long downstream section and the two terminal lakes Lopnor and Taitema completely fell dry. The natural ecosystems along the whole river, especially along the downstream have been degraded severely [19]-[21].

In order to 1) balance the water use between economic development, mainly agriculture, and ecosystem protection and restoration and 2) pursue a more just water distribution between the Aksu River as tributary and the Tarim upper reaches on one side and the Tarim lower reaches on the other side, a water distribution program which gives water quotas for water users and river stretches was adopted by the Xinjiang Government in 2001 [7] [20]. Similar water quota systems had been adopted along the Yellow River [4] and the Heihe River [5].

While a few studies were directed to characterize the institutional development of water management and legal framework of the water distribution program of the Tarim [20] [22] [23], the current reality of the water allocation and effects on water users on the ground has not been systematically assessed yet. Therefore, this paper aims at revealing the water distribution patterns in the Tarim river catchment from a stakeholder perspective by addressing the following four questions: 1) How is water really distributed among different water use sectors, in particular with regard to environmental flow? 2) How is water distributed among different geographic and administration areas? We focus on the Tarim river catchment, because it is more prone to water shortage compared to the Aksu River [13].

The stakeholder perspective was revealed through semi-structured interviews with experts from governmental administrations, river basin management institutions, scientific research institutions, and local farmers.

The following part (Section 2) will introduce the Tarim Basin and the Tarim river catchment regarding its hydrology, climate, land use, and ecological issues. Afterwards, in Section 3 the institutional framework and the water distribution program along the Tarim River will be presented based on a literature review. Section 4 will present the perspectives of the different stakeholders regarding the water distribution in practise. The role of institutions with respect to water distribution is elaborated by interview interpretation and analysis. The paper closes with a discussion and conclusions (Section 5).

2. The Tarim River Catchment

2.1. Geography and Hydrology of the Tarim River Catchment

The Tarim River flows along the northern rim of the Tarim Basin (**Figure 1**). Administratively, the Tarim Basin is shared by 5 prefectures (Bayangol, Aksu, Kizilsu, Kashgar, and Hotan) and four divisions of Xinjiang Production Construction Corps (XPCC). The XPCC is a semi-military governmental organization, which was established in 1954, in order to develop remote frontier regions, to promote land reclamation, and to consolidate border defense [24]. The XPCC administers its divisions independently from the prefectural governments. The XPCC is less dependent from the Xinjiang Government compared to the prefectures.

Until the 1970s, the Aksu, Yarkand, and Hotan River permanently supplied water to the Tarim River. The Tarim started at the confluence of these three tributaries at Aral (**Figure 1**). Today, the Aksu River is the main water source, draining permanently into the Tarim and supplying 73.2% of total amount of the runoff of the Tarim. The Hotan River and the Yarkant River supply 23.2% and 3.6% in average, respectively, and reach Aral only during flood events [6] [7]. The Konqi River (lower part of the Kaidu-Konqi River), which drains the Bosten Lake and which flows parallel to the Tarim lower reaches, was hydrologically connected with the Tarim through natural wetlands and river branches until the 1970s. Today, it supplies water through artificial channels into the Tarim downstream section. The three tributaries of the Tarim as well as all other rivers of the Tarim Basin originate in the mountain ranges, which surround the Tarim Basin, *i.e.* Tianshan, Karakoram, Kunlun and Altun. The central part of the basin is occupied by the Taklamakan Desert [20].

The Aksu, Hotan, Yarkant, and Kaidu-Konqi River as source streams, together with the Tarim mainstream constitute the Tarim river catchment or the so-called *four source streams and one mainstream area* (**Figure 1**). This area is the scope of the Tarim Water Distribution Program. It has a surface of 0.26 million km², accounting 25.4% of the whole Tarim Basin; the total average annual runoff of the four source rivers (including the Konqi) is 25.67 km³, accounting for 64.4% of the whole Tarim Basin [7]. The particular study area of this paper is the Tarim river catchment, *i.e.* the Tarim mainstream sub-basin in **Figure 1**.

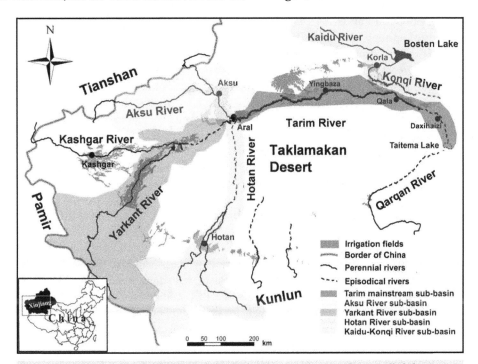

Figure 1. Map of the Tarim Basin and the Tarim river catchment (the so-called *four source streams and one mainstream area*) with the five sub-catchments Aksu, Yarkant, Hotan, Kaidu-Konqi, and Tarim, after [13] and [25].

The climate in the Tarim Basin is extremely arid and continental [26] [27]. The annual precipitation in the Tarim river catchment ranges from 30 to 70 mm, whereas the annual potential evaporation ranges from 2100 to 3000 mm [28]. So, the only water source for natural ecosystems and agriculture is river water. The water supply for the rivers in the Tarim Basin is constituted of melting water from glaciers and snow as well as rainwater from the surrounding mountains. Therefore, the river runoff is concentrated in the period from June to September, amounting to 60% - 80% of the annual runoff, while there is only 10% of the annual runoff from March to May [6] [7] [20].

Once the rivers flow out of the mountains, they turn into losing streams, *i.e.* they do not receive water, but constantly drain water into the groundwater layer and thus become smaller and smaller. The mainstream of the Tarim River is 1321 km long, which is divided into the upper reaches with a length of 495 km (Xiaojiake (close to Aral)—Yingbaza), the middle reaches with 398 km (Yingbaza—Qala), and the lower reaches with 428 km (Qala—Taitema Lake) [6]. Along the middle reaches the Tarim River formed an inland delta, which was about 100 km wide near today's Yingbaza. River branches of this inland delta also connected to the Konqi River [29].

2.2. Economic Development and Ecological Problems in the Tarim River Catchment

The population along the Tarim mainstream has increased from about 0.5 million in 1957 to about 2 million in 2002 [21]. A prognosis provided by [7] states that the population growth rate will be 10.13% from now until the year of 2020. The primary industry (*i.e.* farming, forestry, animal husbandry, and fishery) predominates the economy in the whole Tarim Basin, contributing to 46% of the regional Gross Domestic Product (GDP) [20]. Thereby, cotton agriculture contributes the largest share to the regional GDP. Cotton has become the major crop in the Tarim Basin since the 1990s. While Xinjiang provided only 2.5% of China's total cotton production in 1978, today about one third of China's cotton production comes from Xinjiang [10] [30].

The area of irrigated land increased all over the Tarim Basin, from 706,000 ha in 1949 over 1,330,000 ha in 1980 to 1,412,000 ha in 1990 [31] and 1,650,000 ha in 2008 [8].

Due to the increased irrigation demands, the water resources along the Tarim River have been reduced. Until ten years ago, the water use efficiency of irrigation in the Tarim Basin ranges between 35% and 40% [19]. In 1972, the Daxihaizi Reservoir (**Figure 1**) was constructed. This reservoir cut off the 320 km of the Tarim downstream of Daxihaizi so that this river stretch fell completely dry. The end-lake of the Tarim River, the Lop Nor, comprising an area of 100 km² in the 1950s, disappeared entirely in the early 1970s. The other terminal lake, Taitema Lake; dried up soon afterwards [6] [19]. The salt content of the surface waters and the groundwater has increased due to drainage waters from the irrigation. During the process of land reclamation, farmers and herders already had to give up agricultural land, because they suffered from water shortage and soil salinization [32]. Natural ecosystems have been degraded, especially along the Tarim lower reaches and the rim of the inland delta along the Tarim middle reaches [6].

The natural ecosystems form a mosaic of riparian forests, wetlands and reed beds, and shrub vegetation. The riparian forests consist of *Populus euphratica* and on smaller areas of *P. pruinosa*. The wetlands and reed beds are dominated by *Phragmites australis*. The shrub vegetation consists of *Tamarix* stands and halophytes [15]. Most of the species of those natural ecosystems are so-called phreatphytes. They exploit the groundwater and thus adapt to the arid climate [33], e.g. *P. euphratica* is able to tap the groundwater to depths of 10 m [15].

3. The Water Distribution Program in the Tarim River Catchment

3.1. Water Management at the Tarim River

Until the end of the 20th century, there was no unified water management plan for the Tarim river catchment [20] [34]. The water allocation along all rivers in the Tarim Basin was administered separately by prefectures and divisions of XPCC. Thus, the prefectures and divisions of XPCC competed for water. Within each administrative area, water was diverted according to local demands for irrigation not considering the needs of downstream water users.

Conflicts among administrative areas and individual water consumers, as well as the degradation of the natural ecosystems attracted concerns from the Central Government of China and international institutions. Thus, a comprehensive investigation on water resources and their consumption was carried out in the Tarim Basin during 1975-1982. Since then, an overall plan for a unified water management (UWM) in the entire river basin has been suggested and discussed [20].

In 1992, the Tarim Basin Management Bureau (TMB) was established to manage the overall water resources of the Tarim Basin. Afterwards, in 1997 the Tarim Basin Water Resources Commission (TWRC) was established, in order to guide and supervise the water management in the whole Tarim Basin and Tarim river catchment [35]. From 1991 to 2005, the two World Bank projects *Tarim Basin Project* (1991-1997) and *Tarim Basin II Project* (1998-2005) were carried out, supporting constructions of a more solid infrastructure for irrigation agriculture, increasing efficiency of water and land resource management, and improving the administration structures in the Tarim Basin [36].

In 2001, the State Council adopted the *Near-Future Comprehensive Management Program of the Tarim Basin* [37]. The overarching aim of this program is to protect and partly restore natural ecosystems along the middle and lower reaches of the Tarim River and restore the terminal lake Taitema by guaranteeing a minimum annual amount of water flowing through the Tarim lower reaches downstream of Daxihaizi. A national investment of 10.7 billion CNY (about 1.69 billion US dollars) was allocated for this program from 2001 to 2011. This program focuses on the Tarim river catchment and comprised a bundle of measures: 1) Transfer of water from the neighboring Kaidu-Konqi River catchment into the Tarim lower reaches downstream of the Daxihaizi Reservoir, 2) construction of dykes along the Tarim middle reaches downstream of Yingbaza, in order to channel the water downstream instead of sustaining the inland delta along the middle reaches, 3) propagation of water saving irrigation techniques, and 4) setting up an integrated and fair water distribution program. Two channels from the Konqi River to the Tarim lower reaches were opened by 2004. The lateral dykes along the Tarim middle reaches were completed by 2005. Drip irrigation has been propagated and is used on the majority of cotton fields. The Taitema Lake has reappeared after 30 years varying in size (maximum area was 200 km^2) according to the runoff from the Tarim downstream section. But though this water management program is in place, the Tarim River ceased to flow on more than half of its total river length during spring and early summer 2004, 2007, 2008, and 2009 resulting in crop failures downstream [13].

In 2005, the government of Xinjiang issued two official documents: *Scheme of Surface Water Distribution in the Tarim Basin* (*Four Source Streams and One Mainstream*) and *Management Methods of Water Dispatching in the Tarim Basin* (*Four Source Streams and One Mainstream*). These two documents announced the establishment of the governmental Tarim Water Distribution Program for the Tarim river catchment, which is introduced underneath [38] [39].

3.2. The Tarim Water Distribution Program

The Tarim Water Distribution Program distinguishes three basic water use sectors: production (agricultural and industrial production), domestic water consumption, and environmental flow. Within the sector production irrigated agriculture (mainly cotton) consumes most water. **Figure 2** shows the annual water demands of these three sectors as planned to be implemented in the Tarim Water Distribution Program. These numbers refer to the water demands of the four source steams and one mainstream area, *i.e.* along the rivers Aksu, Yarkant, Hotan, Konqi, and the Tarim proper [7].

Based on the water demands, within the Tarim Water Distribution Program a water quota system has been designed with detailed water quotas for each tributary, the Tarim mainstream, and the administrative areas. First, quotas were made, which determine the amount of annual water release from the tributaries to the Tarim mainstream under average conditions: 3.42 km^3/a stem from the Aksu River, 0.90 km^3/a from the Hotan River, and 0.33 km^3/a from the Yerkant river, which sum up to 4.65 km^3/a at Aral. Additionally, a quota of 0.45 km^3/a from the Kaidu-Konqi River to be released to the Tarim at Qala was fixed (**Table 2** and **Figure 3**). Secondly, water quotas were set for the administration areas. **Table 1** shows the water quotas for Aksu Prefecture and Division 1 of XPCC along the Aksu River. Under average conditions, which means an annual inflow of 8.06 km^3 into the Aksu River, the water consumption along the Aksu River amounts to 4.64 km^3/a. 2.54 km^3/a are diverted into Aksu Prefecture, while 2.11 km^3/a are diverted into Division 1 of XPCC (**Table 1**). Subsequently, 3.42 km^3/a of water are to be released into the Tarim mainstream at the Aral Gouging station (**Table 1**, **Table 2** and **Figure 3**). Similar water quota regulations are also applied for the Yerkant and Hotan River [38].

Water quotas for the Tarim mainstream are more detailed regarding the water use of different sectors. During the period from 1957 to 2000, the maximum annual water inflow at Aral Station was 6.96 km^3/a, while the minimum was 2.56 km^3/a. Therefore, ten water allocation keys were developed (**Table 2**) for the Tarim main-

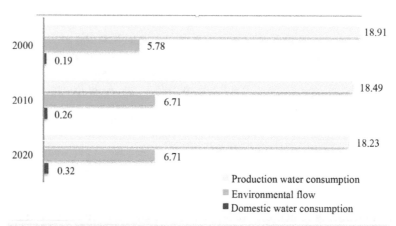

Figure 2. Planned water demands for production (agricultural and industrial production), domestic water consumption, and environmental flow [km³/a] for the Tarim river catchment and the tributaries' catchments (Aksu, Yarkant, Hotan, Konqi) [7].

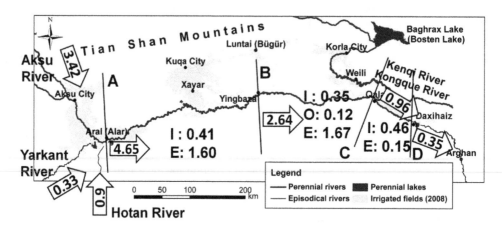

Figure 3. Water inflow from the tributaries Aksu, Yarkant, and Hotan into the Tarim mainstream and water quotas along the Tarim mainstream under average conditions [km³/a], after [13]. Note: I: irrigation and industry, E: environmental flow, O: oil exploitation. A-B: upper reaches, B-C: middle reaches, C-D: upper section of lower reaches, D and below: lower reaches.

Table 1. Annual water quotas within the Tarim Water Distribution Program under average conditions along the Aksu River [km³/a] as the major tributary to the Tarim mainstream [7].

Inflow of headwaters into Aksu River	Water consumption along Aksu River			Water release into Tarim mainstream
	Sum of water consumption	Aksu Prefecture	Division 1 of XPCC	
8.06	4.64	2.54	2.11	3.42

stream varying from an annual inflow of 2.5 km³/a to 7.0 km³/a at Aral. When, for example, an average annual inflow at Aral Station is forecasted (4.65 km³/a), the annual water quota for upper reaches of the Tarim mainstream shall be 2.01 km³/a in total with 0.41 km³/a allocated for production and domestic use, and 1.60 km³/a for environmental flow. The boundary between upper and middle reaches is also the boundary between the Aksu Prefecture and Bayangol Prefecture. Thus, the quota for the upper reaches of the Tarim is equivalent to the quota for Aksu Prefecture along the Tarim River [38].

Under such average conditions, 2.64 km³/a water shall reach Yingbaza and thus shall be released into the middle reaches (**Figure 3**), The annual water quota for the middle reaches until Qala shall be 2.14 km³ in total,

Table 2. Annual water quotas [km³/a] fixed within the Scheme of Surface Water Distribution for water withdrawal for economic activities and environmental flow along the upper, middle, and lower reaches of the Tarim River starting in Aral, i.e. at the confluence of the three tributaries Aksu, Yarkant, and Hotan. The column in light gray refers to the average conditions [7]. The term economic activities refer to irrigation, industrial, and domestic water consumption (*i.e.* sum of production and domestic use).

River section and water user	Annual runoff [km³/a]										
Upper reaches											
Inflow at Aral	2.50	3.00	3.50	4.00	4.50	4.65	5.00	5.50	6.00	6.50	7.00
Economic activities	0.37	0.37	0.37	0.41	0.41	0.41	0.41	0.41	0.41	0.41	0.41
Environmental flow	0.72	0.93	1.15	1.32	1.54	1.60	1.76	1.97	2.19	2.40	2.62
Runoff released into middle reaches	1.42	1.70	1.99	2.27	2.56	2.64	2.84	3.12	3.41	3.69	3.97
Middle reaches											
Runoff at Yingbaza	1.42	1.70	1.99	2.27	2.56	2.64	2.84	3.12	3.41	3.69	3.97
Oil exploitation	0.12	0.12	0.12	0.12	0.12	0.12	0.12	0.12	0.12	0.12	0.12
Other economic activities	0.30	0.30	0.30	0.35	0.35	0.35	0.35	0.35	0.35	0.35	0.35
Environmental flow	0.73	0.95	1.18	1.37	1.60	1.66	1.83	2.06	2.29	2.51	2.74
Runoff released into lower reaches	0.27	0.33	0.38	0.43	0.49	0.51	0.54	0.60	0.65	0.71	0.76
Lower reaches											
Water transferred from Konqi River	0.45	0.45	0.45	0.45	0.45	0.45	0.45	0.45	0.45	0.45	0.45
Runoff at Qala Station	0.72	0.78	0.83	0.88	0.94	0.96	0.99	1.05	1.10	1.16	1.21
Economic activities	0.41	0.41	0.41	0.46	0.46	0.46	0.46	0.46	0.46	0.46	0.46
Environmental flow	0.09	0.10	0.12	0.12	0.14	0.15	0.17	0.21	0.24	0.28	0.31
Runoff released into the lower reaches below Daxihaizi	0.23	0.26	0.30	0.31	0.34	0.35	0.36	0.38	0.40	0.42	0.44

with 0.35 km³, 0.12 km³, and 1.66 km³ allocated for irrigation agriculture, oil industry, and environmental flow, respectively. This river stretch lies within Bayangol Prefecture. Therefore, this water quota is equivalent to the water quota for Bayangol [7] [38].

The remaining 0.51 km³/a, together with 0.45 km³/a carried through channels from the Konqi River, shall be released from the Qala Gauging Station to the lower reaches (**Figure 3**). Along the river stretch from the Qala Gauging Station to the Daxihaizi Reservoir, 0.46 km³/a and 0.15 km³/a are allocated for production (including domestic use) and environmental flow, respectively. Between Qala and Daxihaizi, most of the land lies within the Division 2 of XPCC so that this water quota is equivalent to the quota for the Division 2 of XPCC. Finally, 0.35 km³/a of water shall be released from the Daxihaizi Reservoir into the lower reaches as environmental flow towards the Taitema Lake, which is the current terminal lake of the Tarim River [7] [38] [39].

The Tarim Water Distribution Program grants a constant water amount for agriculture along all sections of the Tarim River. The water amount for agriculture drops slightly only below an annual inflow of 4 km³ at Aral (**Table 2**). A nearly constant amount of water (0.12 km³/a) is guaranteed to the oil exploitation along the Tarim middle reaches over the whole range of annual runoff The water which exceeds the amounts granted to economic activities is released for the natural ecosystems, *i.e.* constitutes the environmental flow. The environmental flow thus varies considerably depending on the inflow into the Tarim River (**Table 2**).

The hydrological year for the Tarim river catchment starts by October and lasts until September of the following year. During winter, an annual inflow for the current hydrological year is estimated. Based on this estimate, annual water quotas for each river stretch and administrative area are set according on **Table 2**. Finally, contracts on these water quotas and management provisions are signed between the representative of TWRC and representative of each administrative area [39].

The hydrological year splits into a so-called non-regulating period, *i.e.* from October to June (which is usually non-flooding seasons) and a so-called regulating period *i.e.* from July to September (which is usually flooding seasons). Right before the beginning of regulating period, the water consumption is calculated, which has been realized by each administrative area so far during the non-regulating period. This amount of water is deducted from the total annual water quota. The remaining water quota is allocated during the regulating period. During the regulating period monitoring and adjusting on water release and diversion are conducted every ten days. During the non-regulating period from October through winter and spring until June no monitoring takes place.

3.3. Institutional Arrangements

According to the Water Law of China [40], water resources are national properties and the state has the responsibility to administer and protect them. Thus, the water resources of the Tarim river catchment are under the overall responsibility of the Xinjiang Government. The TWRC, established as the agency of the regional government, takes full responsibility to administer and protect the water resources in the Tarim river catchment. **Figure 4** shows the decision-making structure of water management in the Tarim River.

The Standing Committee of TWRC is the decision-making body with the following members [41] [42]: Vice-Governor of Xinjiang, representatives of relevant administrations on the provincial level (administrations of water, agriculture and land management, environment, finance and planning), directors of the five prefectures (Bayangol, Aksu, Kizilsu, Kashgar, and Hotan Prefectures) and directors of XPCC's Water Resource Bureau and XPCC's four divisions (Division 1, 2, 3, and 14 of XPCC). It is important to note that the local governments within the XPCC also participate in the management of water resources of the Tarim river catchment, though the XPCC has its own governmental system independent from the Government of Xinjiang, The director of the XPCC's Water Resource Bureau and the directors of the four divisions are commissioners of the TWRC as stated in the Charter of TWRC from 1999 [41].

The TMB, as the administrative and technical agency of TWRC, is the main executive body of the development, utilization, protection, and management of water resources of the Tarim river catchment.

4. Stakeholder Perspectives on the Water Allocation in the Tarim River Catchment

Semi-structured interviews were conducted with 27 exports from governmental administrations (water, agriculture, forest, and pasture at provincial, prefectural, and county levels), river basin management institutions, scientific research institutions and local farmers, in order to investigate the perspectives of different stakeholders with regard to the implementation of the water distribution program. The interviews covered all cities and counties,

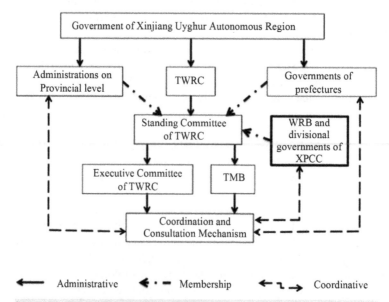

Figure 4. The decision-making structure of water management in the Tarim river catchment, based on the Charter of [20] [41].

which divert water directly from the Tarim mainstream: Aksu City, Aral City, Shaya County, Korla City, Yuli County, and XPCC.

4.1. Stakeholders' Perspectives on the Water Distribution System

The interviews with regard to stakeholders' perspectives on the water distribution system covered the following topics: 1 implementation of the Tarim Water Distribution Program, 2) water distribution between economic activities and environmental flow, 3) water allocation between administrative areas, 4) situation of the natural riparian ecosystems, and 5) evaluation of the water transfer from the Konqi River to the Tarim lower reaches as well as the construction of dykes along the Tarim middle reaches.

With respect to an overall evaluation of the Tarim Water Distribution Program four stakeholders stated that the Tarim Water Distribution Program and its implementation were good, while three stakeholders stated that both were not good. Five stakeholders replied that the Tarim Water Distribution Program was well designed, but its implementation had short comings. The number of respondents to this question was low, because only interviewees from TMB and Xinjiang Water Resource Bureau gave clear statements. All others did not have any clear knowledge about the content of the Tarim Water Distribution Program. Some stakeholders did not reply, because they did actually do not even know that the Tarim Water Distribution Program existed.

In response to the question of water distribution between water use sectors, the opinions were summarized as follows: The TMB, as the top management organization of water resources in the Tarim river catchment and the whole Tarim Basin, stated that the three purposes of water use should follow the order: priority on domestic water use, afterwards water use for production, and the remaining water is left for environmental flow. Theoretically, the water quotas for environmental flow should be guaranteed. But, in practice water first shall be used to meet the demands of all economic activities. The remaining water can be released as environmental flow for natural ecosystems. The opposite opinion was stated by ecologists and forestry officials. They claimed that water use for agriculture and other economic activities should be limited, while environmental flow should be increased.

Responding to the question of water distribution between different administrative areas, five stakeholders agreed that the water distribution between different areas was fair and equal. These stakeholders gave the following reasons: 1) the water quotas were made based on scientific research, 2) all stakeholders agreed on this water distribution plan; and 3) TMB was not a benefit holder but a manager. Five stakeholders replied that the water quotas for different areas in theory were fair, however, the implementation was difficult due to management problems and resulted in periods of water shortage. The two stakeholders Water Resource Bureau of Yuli County (lower reaches) and Aksu Management Bureau (Aksu River) responded that the distribution between the different areas was unfair.

The Water Resource Bureau of Yuli County stated that the water distribution was unfair, because 1) most of the time the lower reaches received less water than their respective water quotas and 2) part of the Konqi River water was diverted to the Tarim River, while in the near past the Konqi River was not a tributary of the Tarim and only flew through Yuli County. This water transfer, which the Water Resource Bureau of Yuli County referred to, was explained in **Figure 3** and **Table 2**.

The Aksu Management Bureau responded, too, that the water allocation between administrative areas was unfair. The reasons given by the Aksu Management Bureau were: 1) about 66,700 ha agriculture land of Aksu Prefecture were not considered in the Water Distribution Program and 2) the water quota for Division 1 of XPCC was separated from Aksu Prefecture, 3) there was no quota for environmental flow in the program for the Aksu River, though Aksu Prefecture had obligations to preserve existing riparian ecosystems, and 4) the development of Aral City may take away water from Aksu Prefecture. Aral became a separate administrative unit in 2009. Before, it was part of division 1 of the XPCC.

Regarding the current situation of the natural ecosystems, the Xinjiang Water Resource Bureau, again, was the only institution, which claimed that the natural ecosystems were recovering all over the Tarim river catchment. However, one stakeholder at the middle reaches and another at the lower reaches claimed that the natural ecosystems inside the dykes were recovering, while the natural ecosystems outside the dykes degraded since the dykes had been constructed. Generally, interviewees from academic institutions and stakeholders located at the lower reaches stated that the natural ecosystems were further degrading.

Regarding the construction of dykes along the Tarim middle reaches, interviewees from academic institutions complained about the necessity of this project and stated that the dykes hampered seed germination of *Populus euphratica*, which was the key-stone species of the riparian forests in the Tarim Basin. Major river branches of the inland delta of the Tarim middle reaches were connected to the Tarim through locks in the dyke. But the locks were opened only for short times so that the river branches carried water for short time periods. The groundwater was refilled only during short times so that the groundwater levels behind the dykes dropped and the vegetation degraded.

Stakeholders from Bayangol Prefecture at the lower reaches of the Tarim River claimed that it was unfair to transfer water from the Konqi River to the Tarim lower reaches, because the Konqi River was not a natural tributary river of the Tarim River anymore. However, Xinjiang Water Resource Bureau, as the only institution holding a positive attitude towards the project, claimed that the water transfer was designed according to the governmental water distribution plan in order to transport water more efficiently and that it was based on scientific research results.

4.2. The Role of Institutions in Water Distribution in Stakeholders' Perspective

Regarding the role of the institutions from the field of water distribution, opinions were collected with respect to three aspects: 1) functions and influences of laws on water distribution; 2) cooperation and interaction between institutions of different levels and from different administration areas; 3) status of public participation in the decision-making process and protection of the rights of local people.

Most interviewees did not reply to the questions about functions and influences of laws on water distribution. One interviewee from the academia pointed out that the current water laws did not entitle people water rights. Moreover, there were no clearly defined punishments for those administrative areas which used more water than granted by the quota system. An interviewee from Xinjiang University thinks that the water distribution program was not implemented according to the distribution plan but according to the person who was implementing it. A number of interviewees expressed their disappointment on the law enforcement.

At the time of the interviews, the Management Bureaus of the tributaries were merged into the TMB. This institutional reform faced many difficulties and the cooperation efficiency was low, because the Management Bureaus of the tributaries used to act rather independently from the TMB. Further conflicts existed between the agriculture and forest administrations, and between institutions from upper reaches and lower reaches. Cooperation between administrative and academic institutions was not well developed. Administrative institutions at local level complained that they had no power to rule affairs within their own administration area, instead they had to report the problems to TMB and wait for replies, which made the management very inefficient.

Most interviewees also did not reply to the question about public participation in the decision-making process and protection of the rights of local people. Those who replied stated that farmers and animal herders had no channel to participate in the decision making processes.

5. Discussion and Conclusions

At the Tarim River and its tributaries a water quota system, the Tarim Water Distribution Program, was set up in a top-down approach, similar to the Yellow River [4] and the Heihe River [5], both in China. In spite of the Tarim Water Distribution Program and its water quotas along the Tarim River, the interview results indicate ongoing conflicts between water using sectors as well as between upstream and downstream.

The water amounts stated in the Tarim Water Distribution Program as listed in **Table 2** and shown in **Figure 3** are annual amounts. Though, the runoff of the Tarim River is concentrated between July and September, while the water demand for irrigation of cotton starts to increase in May when cotton enters the development stage [43] [44]. Thus, the spring time and early summer poses a bottle neck for the water distribution. During this time of the year water users along the Aksu and Tarim upstream withdrawal water from the river according to their demand so that the Tarim may run dry as in 2004, 2007, 2008, and 2009 [13]. Afterwards, during the flood season, in most years there is more water than the irrigation and oil exploitation can use. This water constitutes the largest share of the environmental flow.

During the bottleneck period in spring and early summer, no monitoring of water withdrawal takes place, because this is the so-called non-regulating period. The regulating period with regular monitoring falls into the time, when the Tarim River carries flood. The absence of monitoring of water withdrawal during the bottleneck period in spring and early summer, coupled with partly weak implementation of the Tarim Water Distribution

Program, may explain water allocation below the quotas as stated by water users downstream. Under the current situation, the water users along the middle and lower reaches bear a high risk of losing their crop due to water shortage in early summer. In order to reduce this risk, farmers and villages invest in groundwater wells, which exploit fossil groundwater [13]. Water users with financial and political advantages gain more water security than others. The annual water quotas should be further differentiated into seasonal quotas, which are also monitored during spring, in order to reduce the risk losing their crops and reduce the investment burden for water users downstream. So, the so-called non-regulating period should be abolished.

According to the Tarim Water Distribution Program, first the lower reaches downstream Daxihaizi shall be served at the cost of the middle reaches. But, water shortages outside the dykes along the middle reaches impact on the two nature reserves Tarim Huyanglin and Shaya Talimu Shangyou. Through the dykes and land use, the part of the Tarim Huyanglin Nature Reserve, which still contains near natural riparian ecosystems and which has the potential to protect the riparian ecosystems, including its ecosystem processes, shrunk from 120 km river length to a river stretch of 25 km [15] [45]. In terms of area and in terms of completeness of ecosystem processes, the riparian vegetation preserved along the Tarim lower reaches does not compensate for this impact along the middle reaches.

With respect to institutional settings and public participation, the following suggestions are proposed: 1) The awareness for the significance of ecosystem protection and for the relevant laws and regulations should be increased. 2) Changes should be made in the administrative system, in order to improve the efficiency of institutional cooperation. 3) Efforts should be made to improve the legislation situation. Laws and regulations should be strictly enforced in order to implement the water distribution program. The rights of participation should be guaranteed by law, as it is essential to ensure justice in the field of distribution of public goods. 4) It should be considered to establish a water market. Under a water market, it must be ensured that poor people still have access to water. Water is crucial for all human activities so that water distribution should not favour the advantaged water users.

Acknowledgements

We express our thanks to the Stemmler-Foundation within the German Science Centre, which granted the Master Course Scholarship to Haiyan Peng. Furthermore, we thank the Ministry of Education and Science of Germany as well as the Robert-Bosch-Foundation for providing additional travel funds for the field work.

References

[1]   ICWC (Interstate Commission for Water Coordination of Central Asia) (1992) Agreement of Five Central Asian States. http://www.icwc-aral.uz/statute1.htm

[2]   GLOWA Jordan River (2008) An Integrated Approach to Sustainable Management of Water Resources under Global-Change. http://www.glowa-jordan-river.de/

[3]   Murray Darling Basin Commission (2006) Murray Darling Basin Agreement. http://www.comlaw.gov.au/Details/C2011C00160/Html/Text#_Toc289261269

[4]   Zhu, Z.P., Giordano, M., Cai, X.M., Molden, D., Hong, S.C, Zhang, H.Y., Lian, Y., Li, H., Zhang, X.C., Zhang, X.H. and Xue, Y.P. (2003) Yellow River Comprehensive Assessment: Basin Features and Issues, IWMI and YRCC. http://www.iwmi.cgiar.org/Publications/Working_Papers/working/WOR57.pdf

[5]   Chen, Y., Zhang, D.Q., Sun, Y.B., Liu, X.N., Wang, N.Z. and Savenje, H.H.G. (2005) Water Demand Management: A Case Study of the Heihe River Basin in China. *Integrated Water Resource Assessment*, **30**, 408-419.

[6]   Song, Y.D., Fan, Z.L., Lei Z.D. and Zhang F.W. (2000) Research on Water Resources and Ecology of Tarim River, China. Xinjiang People's Press, Urumqi.

[7]   Tang, D.S. and Deng, M.J. (2010) On the Management of Water Rights in the Tarim Basin. China Water Power Press, Beijing.

[8]   Xinjiang Statistics Bureau (2010) Xinjiang Statistical Yearbook of 2010. China Statistics Press, Beijing.

[9]   FAOSTAT (2013) http://faostat.fao.org/

[10]   USDA (2012) China—Peoples Republic of Cotton and Products Annual. GAIN Report Number: CH12031. http://www.thefarmsite.com/reports/contents/chinacotmay12.pdf

[11]   Feike, T., Mamitimin, Y., Li, L., Abdusalih, N. and Doluschitz, R. Development of Agricultural Land and Water Use and Its Driving Forces in the Aksu-Tarim Basin, P.R. China. *Environmental Earth Science*.

[12] Liu, M.G. (1997) Atlas of Physical Geography of China, China Map Press, Beijing.

[13] Thevs, N. (2011) Water Scarcity and Allocation in the Tarim Basin: Decision Structures and Adaptations on the Local Level. *Journal of Current Chinese Affairs*, **3**, 113-137.

[14] UNESCO (2010) Taklimakan Desert—*Populus euphratica* Forests. http://whc.unesco.org/en/tentativelists/5532/

[15] Thevs, N., Zerbe, S., Peper, J. and Succow, M. (2008) Vegetation and Vegetation Dynamics in the Tarim River Flood-Plain of Continental-Arid Xinjiang, NW China. *Phytocoenologia*, **38**, 65-84. http://dx.doi.org/10.1127/0340-269X/2008/0038-0065

[16] Thevs, N., Zerbe, S., Gahlert, F., Mijit, M. and Succow, M. (2007) Productivity of Reed (*Phragmites australis* Trin. ex Steud.) in Continental-Arid NW China in Relation to Soil, Groundwater, and Land-Use. *Journal of Applied Botany and Food Quality*, **81**, 62-68.

[17] Thevs, N., Buras, A., Zerbe, S., Kühnel, E., Abdusalih, N. and Ovezberdyyeva, A. (2012) Structure and Wood Biomass of Near-Natural Floodplain Forests along the Central Asian Rivers Tarim and Amu Darya. *Forestry*, **81**, 193-202. http://dx.doi.org/10.1093/forestry/cpr056

[18] Wang, S.J., Chen, B.H. and Li, H.Q. (1996) Euphrates Poplar Forest. China Environmental Science Press, Beijing.

[19] Feng, Q., Liu, W., Si, J.H., Su, Y.H., Zhang, Y.W., Cang, Z.Q. and Xi, H.Y. (2005) Environmental Effects of Water Resource Development and Use in the Tarim Basin of Northwestern China. *Environment Geology*, **48**, 202-210. http://dx.doi.org/10.1007/s00254-005-1288-0

[20] Zhang, J.B. (2006) Water Management Issues and Legal Framework Development of the Tarim Basin. In: Wallance, J. and Wouter, P., Eds., *Hydrology and Water Law—Bridging the Gap*: *A Case Study of HELP Basins*, IWA Publishing, London, 108-142.

[21] Hao, X.M., Chen, Y.N. and Li, W.H. (2009) Impact of Anthropogenic Activities on the Hydrologic Characters of the Mainstream of the Tarim River in Xinjiang during the Last Past 50 Years. *Environment Geology*, **57**, 435-445. http://dx.doi.org/10.1007/s00254-008-1314-0

[22] Wang, R. (2009) An Empirical Analysis of Development of Water Rights in China. *US-China Law Review*, **6**, 39-49.

[23] Lyle, C. and Mu, G.F. (2011) Integrated Management of the Tarim Basin, Xinjiang China. http://www.watertech.cn/english/clivelyle.pdf

[24] XPCC (2007) A Brief Introduction of XPCC. http://www.xjbt.gov.cn/publish/portal0/tab130/info25.htm

[25] TMB (Tarim Management Bureau) (2011) Report about the Reform of the Administration System of the Unified Management of the Tarim River Catchment. http://tahe.gov.cn/zhuanti/gaige/banner.html

[26] Tang, Q.C., Qu Y.G. and Zhou, L.C. (1993) Hydrology and Water Resource Utilization of Arid Land in China. China Science Press, Beijing.

[27] Zhou, L.C. (1999) River Hydrology and Water Resources of Xinjiang. Xinjiang Scientific Sanitation Press, Urumqi.

[28] Feng, Q. and Cheng, G.D. (1998) Current Situation, Problem and Rational Utilization of Water Resources in Arid North-Western China. *Journal of Arid Environment*, **40**, 373-382. http://dx.doi.org/10.1006/jare.1998.0456

[29] Hedin, S.A. (1903) Scientific Results of a Journey in Central Asia 1899-1902, Vol. 1, The Tarim River. Lithografic Institute of the General Staff of the Swedish Army, Stockholm.

[30] Hsu, H.H. and Gale, F. (2001) Regional Shifts in China's Cotton Production and Use. Economic Research Service/USDA, Cotton and Wool Outlook/CWS-2001, 19-25.

[31] Xia, D.K. (1998) Dynamics and Water Resources of the Tarim River in Xinjiang. *Journal of Arid Land Resources and Environment*, **12**, 7-14.

[32] Hoppe, T. (1992) Chinesische Agrarpolitik und Uygurische Agrarkultur im Widerstreit. Das Sozio-Kulturelle Umfeld von Bodenversalzungen und-Alkalisierungen im Nördlichen Tarim-Becken (Xinjiang). Institut für Asienkunde, Hamburg.

[33] Gries, D., Zeng, F., Foetzki, A., Arndt, S.K., Bruelheide, H., Thomas, F.M., Zhang X.M. and Runge, M. (2003) Growth and Water Relation of *Tamarix ramosissima* and *Populus euphratica* on Taklamakan Desert Dunes in Relation to Depth to a Permanent Water Table. *Plant Cell Environment*, **26**, 725-736. http://dx.doi.org/10.1046/j.1365-3040.2003.01009.x

[34] Mao, X.H. (2001) Study on Sustainable Utilization Strategy of Water Resources in the Tarim Basin. *Arid Land Geography*, **24**, 136-140.

[35] TMB (Tarim Management Bureau) (1997) Regulations of Tarim Basin on Water Resources Management. http://www.tahe.gov.cn/e/action/ShowInfo.php?classid=106&id=4963

[36] World Bank (2007) Tarim Basin II Project. Report No. 41122. Sector, Thematic, and Global Evaluation Division Independent Evaluation Group, World Bank, Washington, 45-83.

[37] TMB (Tarim Management Bureau) (2001) Report of Near-Future Comprehensive Management Program of the Tarim Basin. http://www.tahe.gov.cn

[38] TMB (Tarim Management Bureau) (2005) Scheme of Surface Water Distribution in the Tarim Basin (Four Source Streams and One Mainstream). http://www.tahe.gov.cn/e/action/ShowInfo.php?classid=106&id=7375

[39] TMB (Tarim Management Bureau) (2005) Management Methods of Water Dispatching in the Tarim Basin (Four Source Streams and One Mainstream). http://www.tahe.gov.cn/e/action/ShowInfo.php?classid=106&id=7377

[40] Water Law of the Peoples Republic of China. http://www.chinawater.net.cn/law/waterlaw.htm

[41] TMB (Tarim Management Bureau) (1999) Charter of the Tarim Basin Water Resources Commission. http://www.tahe.gov.cn

[42] Radosevich, G.E. (1999) Tarim Basin Case Study. IUCN Water Program NEGOTIATE Case Study. http://cmsdata.iucn.org/downloads/china_1.pdf

[43] Hofmann, S. (2006) Comparative Analysis of Uyghur and Han-Chinese Farm Management along the Middle Reaches of the Tarim River. In: Hoppe, T., Kleinschmit, B., Roberts, B., Thevs, N. and Halik, Ü., Eds., *Watershed and Floodplain Management along the Tarim River in China's Arid Northwest*, Shaker, Aachen, 359-371.

[44] Bothe, J. (2010) The Water Use of Cotton Cultivation in the Tarim Basin in Northwest China. Bachelor Thesis, University of Osnabrück, Osnabrück.

[45] Thevs, N., Zerbe, S., Schnittler, M., Abdusalih, N. and Succow, M. (2008) Structure, Reproduction and Flood-Induced Dynamics of Riparian Tugai Forests at the Tarim River in Xinjiang, NW China. *Forestry*, **81**, 45-57. http://dx.doi.org/10.1093/forestry/cpm043.

2

Adaptation Technology: Benefits of Hydrological Services—Watershed Management in Semi-Arid Region of India

Anupam Khajuria*, Sayaka Yoshikawa, Shinjiro Kanae

Department of Civil Engineering, Tokyo Institute of Technology, Tokyo, Japan
Email: *khajuria.a.aa@m.titech.ac.jp

Abstract

Watershed management consists of multifunctional activities to manage and address the increasing water resource problems. Ever increasing water demand and rapidly depleting water resources, it has become necessary to develop the adaptation options to recharge groundwater resources. A watershed is a special kind of Common Pool Resources (CPRs); an area is defined by hydrological linkages where optimal management requires coordinating the use of natural resources by public participation. Watershed developments have shown significant positive impacts on water table, perennially of water in wells and water availability especially in semi-arid regions. This paper describes direct and indirect impacts of the watershed activities and benefits of hydrological services dealing with watershed management with future prediction of net irrigation water supply. In the present work, we have also discussed the multiple impacts of watershed of CPRs for improving groundwater and surface water resources.

Keywords

Watershed Development, Adaptation Options, Hydrological Services, Ground Water, Common Pool Resources, India

1. Introduction

Climate change is projected to have significant impacts on water resources affecting agriculture. It may lead to an intensification of extreme of the global hydrological cycle and could have major impacts on both groundwater and surface water supply [1]. Groundwater resources play a major role in ensuring livelihood security across

---

*Corresponding author.

the world, especially in economies that depend on agriculture. The efficient use of water as resource conservation, environmental friendliness, appropriateness of technology, economic viability and social acceptability of development issues are major priority for agriculture in water scarce regions [2].

Adaptation refers to "adjustments in ecological, social or economic systems in response to actual or expected climatic stimuli and their effects or impacts. It refers to changes in processes, practices and structures to moderate potential damages or to benefit from opportunities associated with climate change" [3]. The possible adaptation responses to the impact of irrigation water resources include both supply and demand side [4]. Adaptation option "Watershed technology" is one of a way to raise rainfed agricultural production, conserve natural resources and increase water availability in agricultural purposes especially in the world's semi-arid and tropical regions. Watershed development is also a special case of multiple-use of Common Pool Resources (CPRs) by showing the linkages of optimize use of groundwater for conservation, agricultural productivity and benefit-cost of all users [5]. And also watershed management is a best approach to managing forestry, agriculture, pasture and water management, with an objective of sustainable management of natural resources. This article discusses the role and impact of watershed management and investigates the benefits of hydrological services of effective and properly managed watershed programme in India. In this article, we choose, discuss and determine the impacts of watershed challenge of the case study of "Gokulpura-Goverdhanpura watershed" semi-arid region of India. And also we assess the multi-faced benefits "Common Pool Resources" of watershed development.

2. Watershed Development and Its Role and Impacts

Watershed development is the process of organizing the use of natural resources to provide necessary goods and services to people, while mitigating the detrimental impacts of land-use activities on soil and water resources. This approach recognizes the intrinsic interrelationships among soil, water, and land use and the connections between upland watersheds and larger downstream river basins.

Watersheds provide a diversity of benefits to local inhabitants and to a greater number of people within the larger river basin through the flows of water and other natural resources off the watersheds. The components of watershed development and its role are summarized in **Table 1**.

The watershed development involving the entire community and natural resources influence 1) productivity and production of crops, changes in land use and cropping pattern, 2) attitude of the community and their participation, 3) socio-economic conditions such as income, employment, assets, health and education, 4) impact on environment, 5) use of land, water, human and livestock resources, 6) development of institutions for implementation and 7) ensuring sustainability of improvement. It is thus clear that watershed development is a key to sustainable production of food, fodder, and fuel wood and meaningfully addresses the social, economic and cultural status of the rural community [6] [7].

The intended impacts of watershed development are to increase groundwater recharge and increase overall water resource availability. Soil and water conservation measures, drainage-line treatments such as check-dams and tree planting are aim to reduce runoff and increase percolation. Watershed development projects may actually contribute to increased competition between water use for irrigation and domestic use, because extending

Table 1. Components of watershed developemnt and its role.

Activity	Objective	Impact
Check dams	Stop/slow down water runoff in gullies	Recharge of groundwater and nearby wells. Creations of open water bodies
Ponds	Groundwater recharge water for cattle	Recharge of groundwater. Creation of big open water bodies
Gully plugs, Gabions	Primarily to trap sediment/silt in gullies and to stabilize the guilty	Keeps sediment out of downstream areas. Increased water infiltration due to slowing down water
Earthen loose boulders	Stabilize existing drainage system	Reduced erosion
Water harvesting plus irrigation	Ensure first and second crops by means of irrigation.	Considerable increase in food production. Increased biomass. Creation of open water bodies. Less water down-stream in catchments
Contour/boundary trenches/bunds	Stop runoff and sediment out of farmer's field	Increased water availability and increased agricultural production. Reduced loss of especially fertile soils

the irrigated area is often an explicit objective or an unintended outcome.

1) Increase in groundwater level: Various factors are accountable for increase in ground water. Land development activities such as contour bunding, land levelling and cultivation practices also contribute towards accumulation of ground water.

2) Increase in surface and stream flow: Both surface water and stream flow has increased during the post watershed development. In Rajasthan, 49% watersheds increase <20% in surface water.

3) Soil erosion reduction: The best performing watersheds are those where soil erosion was reduced by more than 50 percent and the worst performing are the ones where there is an increase in soil erosion or the implementation failed in arresting soil erosion.

4) Runoff reduction: According to the beneficiaries this has been possible because of the contour bunding or field bunding which has also helped in checking the runoff of rainwater resulting in soil moisture retention.

5) Change in land use pattern: With increase in surface water conservation and increase in availability of water in the watershed regions, it is expected that there will be more positive change in land use pattern.

6) Cropping pattern and agricultural productivity: Both agricultural diversification and intensification lead to increase in agricultural productivity in the regions where watershed development are effective.

3. Watershed Development in India

Watershed management has come into focus in India with the advent of crop productivity fluctuations with the green revolution since 1980s. Watershed development is an ever-popular rural development mantra in India. The number and range of programmes continue to increase and the Government of India have invested over US$500 million per year into the rehabilitation of watersheds. Increased emphasis on watershed development programs for dry land plain regions, *inter alia*, is a manifestation of the shifting priorities in the agricultural sector, which until recently concentrated mainly on crops and regions with assured irrigation [8] [9].

The direct and indirect impacts of water services in watershed interventions are summarized in **Table 2**. Irrigation and agriculture were boosted through watershed interventions and the irrigation water use especially arid and semi-arid regions [7] [10].

4. Case Study of Gokulpura-Goverdhanpura Watershed Development

Location: The study was undertaken in the Gokulpura-Goverdhanpura watershed situated in southeastern part of Rajasthan in India. The watershed is located about 40 Km northwest of Blundi town and lies between latitude 25°35'N and longitude 75°25'E. (**Figure 1**) with total area of 1355 ha. The adjoining states are Punjab and Haryana in the North, Uttar Pradesh in the Northeast, Madhya Pradesh in the Southeast, and Gujarat in the Southwest. Rajasthan is the largest State in the country in terms of area and also one with the highest proportion of land occupied by desert. It receives only 1 percent of the available water, which supports 5 per cent of population and 10 per cent of the geographical area [11]. The mean annual rainfall is 420 mm with extreme variation [12]. The development of the Gokulpura-Goverdhanpura watershed started in 1996. Data for the empirical study was collected from several literature of the state of Rajasthan through the International Crops Research Institute for the Semi-Arid Tropics (ICRISAT) research team. The summarized organized 20 structures of check dams have been constructed benefiting the irrigation area about 92.5 ha. The constructed structures have objec-

Table 2. Direct and Indirect impacts of water services.

Watershed site	Direct impacts	Indirect impacts
Govardhanpra-Gokulpura, Rajasthan	-Pit composting to promote recycle agro and domestic waste.	-Water harvesting and recharge measures; -Improved surface water, storage and groundwater levels.
Adihali-Mylanhalli, Karnataka	-New bore well provided for defunct water supply scheme.	-Network of farm ponds; -Improved groundwater recharge and raised water levels.
Karaoia-Sengur-Jamuna, Uttar Pradesh	-Newly developed dug wells and tube wells.	-Conservation measures, augment springs and dug wells.
Kelghar-ranjanpad, Maharashtra	-Community bathing facilities provided.	-Benefit for groundwater sources close to stream.
Kharachiya-Kharahiya jam, Gujarat	-Bore well, pumps, overhead tank and stand pots are Provided.	-Check dams and well; -recharging structures; -Improved groundwater levels until drought (tankers still required).

Figure 1. Location of study area Gokulpura-Goverdhanpura watershed, Rajasthan, India.

tive and impact as shown in above **Table 1**.

Impacts on socio-economic, technical and ecological challenges: Socio-economic challenges: The socio-economic indicators are changes in household income, per capita income, consumption expenditure, employment, migration, people's participation and household assets at the village level were considered [13]. The results on the poverty indicators of watershed development indicate the farmers belonging to marginal and small land holding are getting more benefits. The percentage of per capita income was higher from 8% to 24% [12].

Technical challenges: The structures like check dams, sunken ponds help enhance the surface water storage capacity which helps in improving groundwater recharge. The major significant is increase in groundwater level in wells even during the low rainfall years [14].

Ecological challenges: The area under irrigation increased by 66% after the implementation of the watershed development. This resulted in marked reduction in crop failures and increased farmers 'confidence to invest in improved agricultural inputs [12]. In addition, about 35 ha land was brought under horticulture with irrigation facility. The annual runoff is reduced by 52% and soil loss by 64% and the land-use pattern is changing with watershed constructed structures [12] [14].

5. Benefits of Hydrological Services

Of the many ecosystem services that watersheds provide, hydrological services constitute some of the most economically and socially valuable [15]. These services largely fall into four broad categories: water filtration/purification; seasonal flow regulation; erosion and sediment control and habitat preservation. The increasing vulnerability of agricultural output to variations in rainfall, particularly during droughts when the soil moisture is scarce, is attributed to inadequate expansion of irrigation. Watersheds with a high proportion of land covered by intact forests and wetlands are particularly effective at moderating runoff and purifying water supplies. The vegetation and soils of forests and wetlands in watershed varies with seasonal variation. Even the cost and benefit are distributed unevenly, which is resulting from spatial variation and multiple, conflicting uses of natural resources. The conflict between upper watershed which is using for grazing and down watershed helps to protecting for regeneration to support downstream irrigation. Deforestation, road construction, clear-cutting, and poor farming practices can send large influxes of eroded sediments into rivers and streams, markedly degrading the quality of water and of aquatic habitats [9]. Watershed development activities have significant impact on groundwater recharge, access to groundwater and the expansion in irrigated area. And also these activities have been found alter crop pattern, increase crop yields and crop diversification and enhanced employment and farm income. Watershed protection has reduced capital, operation and maintenance costs in developing countries and contributes to the natural resource conservation.

In order to estimate the agricultural water withdrawal, we have utilized a model based on an integrated global water resource, known as the H08 model [8]. The net irrigation water from each supply source is shown in **Figure 2**. The river and large reservoir are continuously same trend from previous to predicted future years whereas additional water to the total water supply increases.

6. Incentives for Common Pool Resource Management in Watershed Development

Common pool resources (CPRs) are significant for poor people's livelihood with the diverse resources e.g. water, forest and fodder. According to Ostrom (1990) [15], "Whenever one person cannot be excluded from the benefits that others provide, each person is motivated not to contribute to the joint effort, but to free ride on the efforts of the others. If all participants choose to free ride, the collective benefit will not be produced". CPRs are the particular challenge for watershed which is commonly involved into groundwater and surface water to ensure sustainable management of natural resources [16]. In the context of watershed development, CPRs in which the right uses are commonly held in both groundwater and surface water. The watershed development experiences can largely attribute to communities that make sustainable institutional and social issues. The four categories of factors play an important role; a) Resource system characteristics, b) User group characteristics, c) Institutional arrangements, d) External environment. Common pool resources make up large areas within the watershed and their sustainable management.

7. Lesson Learnt and Conclusions

The benefits and future direction of watershed development are described below.

Upstream and downstream flow: As the watershed development structures generate various positive externalities, quantifying the benefits from the structures like check dams, gabion structure and gully plugs. When quantifying the benefits, the zone of influence of check dams varies from 300 m to 400 m downstream and 200 m to 250 m upstream. Similarly, the zone of influence of groundwater recharge varies from 4 km to 5 km downstream [2].

Natural and artificial recharge: The structures like check dams, gabion, gully plugs, sunken ponds and field bunds are expected to increase the groundwater recharge in the wells around 30 percent. However, the natural recharge without watershed structures is reported to be about 10 percent. Thus, the impact of watershed technology care should be taken to account for the natural and artificial recharge.

Policy implication: The study has revealed that watershed developments have significant impacts on groundwater recharge, access to groundwater and the expansion in irrigated area [17] [18]. The policy must be focus on

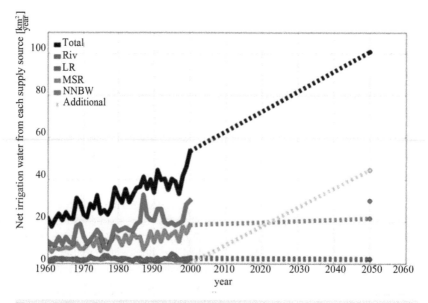

Figure 2. Net irrigation water from each source of water supply in studied area Gokulpura-Goverdhanpura watershed, Rajasthan, India.

the development of structures which are feasible for semi-arid regions.

A coordinated effort is now required to promote watershed development and protect domestic water supplies at the same time. The watershed development projects increase the availability of groundwater at a local scale, while on the other hand, watershed programme frequently lead to greater irrigation water use for scarce water resource.

In conclusion, we have found that the watershed development is relatively feasible for the semi-arid region. There are number of options that can help to facilitate integration of adaptation technology and policy. As future tasks, we should determine deeply other adaptation options for a better and positive impact of climate change. In addition, the public and private participation may be encouraged in a big way to harvest the groundwater recharge and provide employment and farm income.

References

[1] Mimi, A.A. and Jamous, S.A. (2010) Climate Change and Agricultural Water Demand: Impacts and Adaptations. *African Journal of Environmental Science and Technology*, **4**, 183-191.

[2] Pereira, L.S., Oweis, T. and Zairi, A. (2002) Irrigation Management under Water Scarcity. *Agricultural Water Management*, **57**, 175-206. http://dx.doi.org/10.1016/S0378-3774(02)00075-6

[3] IPCC (2007) Climate Change 2007: Synthesis Report. Contribution of Working Groups I, II and III to the Fourth Assessment. *Report of Intergovernmental Panel on Climate Change*, Geneva. http://www.ipcc.ch/ipccreports/ar4-syr.html

[4] Smakhtin, V. (2004) A Pilot Global Assessment of Environmental Water Requirements and Scarcity. *International Water Resources Association*, **29**, 307-317.

[5] Kerr, J. (2007) Watershed Management: Lessons from Common Property Theory. *International Journal of the Commons*, **1**, 89-109.

[6] Shah, A. (1998) Watershed Development Programmes in India: Emerging Issues for Environment-Development Perspectives. *Economic and Political Weekly*, **33**, A66-A79.

[7] Khajuria, A., Yoshikawa, S. and Kanae, S. (2013) Watershed Management: Participatory Issues for Sustainable Livelihood. *The Sixth Conference of the Asia Pacific Association of Hydrology and Water Resources "Climate Change and Water Security—APHW* 2013", Seoul, 19-21 August 2013.

[8] Khajuria, A., Yoshikawa, S. and Kanae, S. (2013) Estimation and Prediction of Water Availability and Water Withdrawal in India. *Annual Journal of Hydraulic Engineering, Japan Society of Civil Engineering, (SUIKO)*, **57**, 145-150.

[9] Badal, P.S., Kumar, P. and Bisaria, G. (2006) Dimensions and Determinants of People's Participation in Watershed Development Programmes in Rajasthan. *Agricultural Economics Research Review*, **19**, 57-69.

[10] Prabhakar, P., Chourasia, A.K., Wani, S.P. and Sudi, R. (2013) Multiple Impact of Integrated Watershed Management in Low Rainfall Semi-Arid Region: A Case Study from Eastern Rajasthan, India. *Journal of Water Resource and Protection*, **5**, 27-36. http://dx.doi.org/10.4236/jwarp.2013.51004

[11] Sharma, R.J.P., Singh, P. and Padaria, R.N. (2011) Social Processes and People's Participation in Watershed Development. *Journal of Community Mobilization and Sustainable Development*, **6**, 168-173.

[12] Pathak, P., Chourasi, A.K., Wani, A.P. and Sudi, R. (2013) Multiple Impact of Integrated Watershed Management in Low Rainfall Semi-Arid Region: A Case Study from Eastern Rajasthan, India. *Journal of Water Resource and Protection (JWARP)*, **5**, 27-36. http://dx.doi.org/10.4236/jwarp.2013.51004

[13] Deshpande, R.S. and Rajasekaran, N. (1997) Impact of Watershed Development Programme: Experiences and Issus. *Artha Vijnana*, **39**, 374-390.

[14] Pathak, P., Wani, S.P., Sudi, R., Chourasia, A.K., Singh, S.N. and Rao, K. (2007) Rural Prosperity through Integrated Watershed Management: A Case Study of Gokulpura-Goverdhanpura in Eastern Rajasthan. *Global Theme on Agroecosystems Report no. 36. Patancheru 502 324, Andhra Pradesh, India: International Crops Research Institute for the Semi-Arid Tropics (ICRISAT)*, **52**, 1-52.

[15] Raju, Sharma, J.P., Singh, P. and Padaria, R.N. (2011) Social Processes and People's Participation in Watershed Development. *Journal of Community Mobilization and Sustainable Development*, **6**, 168-173.

[16] Ostrom, E. (1990) Governing the Commons: The Evolution of Institutions for Collective Action. Cambridge University Press, New York. http://dx.doi.org/10.1017/CBO9780511807763

[17] World Bank (2007) Watershed Management Approaches, Policies and Operations: Lessons for Scaling-Up. Agriculture and Rural Development Department, World Bank, Washington DC.

[18] Swami, V.A. and Kulkarni, S.S. (2011) Watershed Management—A Means of Sustainable Development—A Case Study. *International Journal of Environmental Science and Technology (IJEST)*, **3**, 2105-2112.

3

Comparison and Assessment of Success of Models in Watershed Simulation and Management

Maisa'a W. Shammout

Water, Energy and Environment Center, The University of Jordan, Amman, Jordan
Email: maisa_shammout@hotmail.com, m.shammout@ju.edu.jo

Abstract

In Jordan, Zarqa River Basin (ZRB) has been taken as a case study for applying water management models because of its limited water resources and due to the fact that the basin is dwelling with about 52% of Jordan's population. The surface water resources are mainly used for agriculture because they are mixed with treated water and cannot be used for domestic purposes. This paper aims to demonstrate the contributions of Models in watershed management that provide indirect ways of assessing and confirming the success of models in water flow simulation. The method includes transferring the computed hydrologic parameters for Zarqa basin's sub-catchments within Watershed Modeling System (WMS) into Water Resources Model (WRM) and HEC-1 models. Then the results of the HEC-1 and WRM models are compared according to their basin's simulation with the real basin. The study includes description of the HEC-1, WRM models philosophy, the models representation, and simulation results and analysis of the Zarqa River Basin. Comparing the results of WRM and HEC-1 models proved their simulation efficiency in predicting the flow of Zarqa River Basin. Nevertheless, the philosophy of HEC-1 is a single storm event and is based on values of curve number, while WRM philosophy describes the water flow and availability, and demand and supply balance on a daily basis across the basin. The models' predictions for the real flow definitely establish the modeling certainty and help the water resources' developers to incorporate different basin features for watershed representation, simulation, and management. Hence, the certainty of the results in modeling provides indirect ways of assessing the success of models' simulations.

Keywords

Water Scarcity, Watershed Management, Watershed Modeling System, Water Resources Model, Watershed Simulation

1. Introduction

Jordan is classified among countries of the world with limited water resources and it is considered as one of the lowest on a per capita basis. The available water resources per capita are falling as a result of population growth and are projected to fall from less than 150 m^3/capita/year at present to about 90 m^3/capita/year by 2025, putting Jordan in the category of an absolute water shortage. The scarcity of water in Jordan is the single most important constrains to the country growth and development because water is not only considered a factor for food production but a very crucial factor of health, survival and social and economical development. As a result of scarcity, the demands and uses of water are far exceeding renewable supply. Groundwater that used for domestic purposes are not enough to satisfy the growing demand for domestic and industrial sectors [1].

Surface water resources in Jordan are distributed among 15 major basins. Zarqa River Basin is considered as one of the most significant basin with respect to its economical, social and agricultural importance since it is the second main tributary to Jordan River. Zarqa River basin drains an area of about 4120 square kilometres (km^2) where about 95% of its area is within Jordan and about 5% is in Syria. The basin suffers several water problems ranging from management, quality and conflict issues as well as allocation among sectors [2] [3]. The water system in the basin is a complex characteristics where surface water mixed with treated effluent are stored at a reservoir located at the outlet of the basin and is used for irrigation in the Jordan valley; groundwater within the basin are used for the three sectors; surface fresh water are pumped to the basin from the Jordan valley for domestic use; and ground water from different basins are transferred to meet the growing domestic demand [4]. The heavy utilization of Zarqa basin has resulted in reducing the base flow of Zarqa River from 5 m^3/s to less than 1 m^3/s and the discharge of springs from an average 317 MCM prior to 1985 to less than 130 MCM after 2000. Similar conditions are observed in other side wadis. Moreover, about 52% (3.25 million) of Jordan population lives in the basin compared to 6.249 million in the country [5]. Administratively, the basin is located in five governorates, namely; Amman, Balqa'a, Zarqa, Jarash, and Mafraq. **Table 1** shows these governorates and their populations.

For these reasons, simulating models for watershed management are needed to represent and simulate the critical situation of Zarqa River Basin. Hydrological data are incorporated into the Watershed Modeling System (WMS) represented in HEC-1 model; the rainfall-runoff simulation, and the WaterWare water resources management information system is represented by using Water Resources Model (WRM); the dynamic water allocation and budget Model. HEC-1 model is developed by the Hydrologic Engineering Centre while the WRM is implemented as web-based distributed client-server systems and provided by Environmental Software Systems-ESS, Austria. The results of the HEC-1 and WRM models are compared according to their basin's simulation. The study includes description of the HEC-1, WRM models philosophy, the models representation, simulation results and analysis of the Zarqa River Basin.

This paper aims to demonstrate the contributions of Models in watershed management that provide indirect ways of assessing and confirming the success of models in water flow simulation.

2. Modeling Tools Philosophy

2.1. HEC-1 Model Philosophy

The HEC-1 model is fully linked with the Watershed Modeling System (WMS) Software. HEC-1 is developed by the Hydrologic Engineering Centre. It is the most commonly-used lumped parameter model available. HEC-1 model is designed to simulate the surface runoff response of a river basin to precipitation by representing the ba-

Table 1. Zarqa River Basin Population of the Year 2011.

Governorate	Female	Male	Population
Amman	902,762	963,846	1,866,608
Zarqa	442,043	473,095	915,138
Mafraq	35,959	39,389	75,348
Balqa'a	112,078	119,299	231,377
Jerash	76,489	81,309	157,798
Total	1,569,331	1,676,938	3,246,269

sin as an interconnected system of hydrologic and hydraulic components. It is a single storm event, and includes several options for modeling and computing total runoff volume of each rain storm.

Precipitation loss is calculated based on supplied values of curve number (CN) and initial surface moisture storage capacity. CN and surface moisture storage capacity are related to a total runoff depth for a storm. Curve numbers in hydrology are used to determine how much rainfall infiltrates and how much becomes runoff.

Each component models an aspect of the precipitation-runoff process within a portion of the basin, commonly referred to as a sub-basin. A component may represent a surface runoff entity, a stream channel, or a reservoir. Representation of a component requires a set of parameters which specify the particular characteristics of the component and mathematical relations which describe the physical processes. The result of the modeling process is the computation of stream flow hydrographs at desired locations in the river basin [6].

Watershed Modeling System (WMS) Software supports several techniques to be efficient in preparing the initial model requirements. It is used to define watershed characteristics using a Digital Elevation Model (DEM). WMS is used in determining the flow direction and accumulation, stream network, basin outlet, and interior outlets, for creating the entire watershed and its sub-basins using Topographic Parameterization Program (TOPAZ). The average rainfall, area/elevation can be derived for each sub-basin based on hydrologic sub-basin parameters computed by WMS. The results obtained from WMS are used as an input for models; HEC-1 Model, and the Water Resources Model (WRM).

2.2. Water Resources Management Model Philosophy

The water resources model (WRM) is one of the core components of the Water Ware system. It describes the water flow and availability, demand and supply balance on a daily basis across the basin and its elements, based on conservation and continuity laws [7].

In order to simulate the behaviour of a river basin over time, the river basin is described as a system of nodes and arcs. These nodes represent the different components of a river system as diversions, irrigation areas, reservoirs, etc., and can indicate points of water inflow to the basin, storage facilities, control structures, demand for specific uses. The nodes are connected by arcs which represent natural or man-made channels which carry flow through the river system [8] [9].

3. The Models Representation of Zarqa River Basin

Any year can be entered into the models and thus will be hydrologically analyzed. To represent Zarqa River Basin, the hydrological year of 2001/2002 is taken because it is an average rainy season with good distribution of rainfall amounts. For that year, all data related to groundwater and wells extraction are available and their records are on daily basis. Other data that are also available concern the flood records at one gauging station, metrological data from 10 stations, and influent and effluent of the four wastewater treatment plants (WWTP).

3.1. Stream Network, Sub-Basins Representation via Watershed Modeling System and HEC-1 Model

Zarqa River Basin (ZRB) is subdivided into an interconnected system of stream network components using topographic maps of 1:50,000 scale and Digital Elevation Model (DEM) with resolution of 30 m. **Figure 1** shows the Zarqa River course on topographic map and DEM. The basin is delineated to determine its boundaries, drainage network, and outlet. The basin is divided into 6 sub-basins (SB), namely; Upper Dhleil, Lower Dhleil, Amman, Jerash, Suileh/Ketteh, and Al-Salt sub-basin as shown in **Figure 2**. These sub-basins are subdivided based on basin slope (BS), rainfall amounts and distribution, urban centres, and hydrological boundaries. In each sub-basin, the catchments contribution to the main stream is determined using the weighted average of the sub-basin according to the rainfall distribution and the hydrological characteristics of each sub-basin to the whole basin. **Figure 3** shows the Gages Weights of sub-basin. HEC-1 model within WMS software is run and is verified according to the records of Runoff Gauge at Jerash Bridge (control gauge).

3.2. Stream Network, Sub-Basins Representation via Waterware Applying Water Resources Model

The derived data from WMS are used as input files and initial requirements for running the WRM model. Iden-

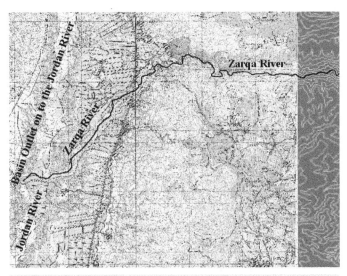

Figure 1. Zarqa River course on the Topographic Map and DEM.

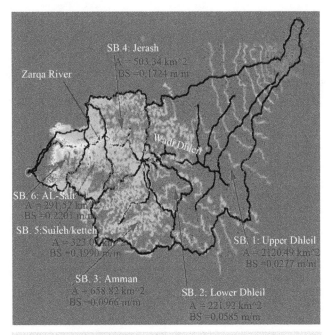

Figure 2. Zarqa River sub-basins.

tification of Zarqa River Basin, sub-basins, streams, and the computed hydrological parameters help in representing and simulating ZRB in terms of nodes and reaches according to facilities provided by Environmental Software Systems-ESS [10] [11]. Thus, all objects of each sub-basin (6-SBs) are displayed on a topology map. The detailed methodology and results of application of WRM for ZRB have been presented in previous papers [4] [12]. **Figure 4** shows the displayed nodes. Supply and demand nodes for different purposes namely; agriculture, domestic, and industry data are specified for functional nodes.

The water resource system in Zarqa River Basin is complicated due to the high water demand compared to the limited water supply. In addition, the surface water resources are mainly used for agriculture because they are polluted with treated water and cannot be used for domestic purposes. The total number of groundwater wells in the basin for the studied year is 752 wells, where 740 of them are productive wells. Data for the daily discharges, the coordinates, and the purpose of use of these wells are obtained from the Ministry of Water and Irrigation [13]. Groundwater that is used for domestic purposes is not enough to satisfy the growing demand of domestic and industrial sectors. Therefore, additional water supply is transferred from other basins namely; Azraq, Lajoun,

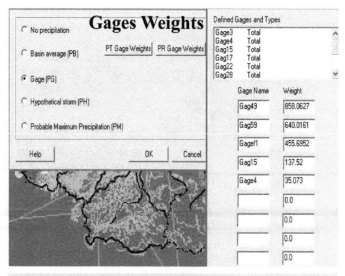

Figure 3. The gages weights of the sub-basins.

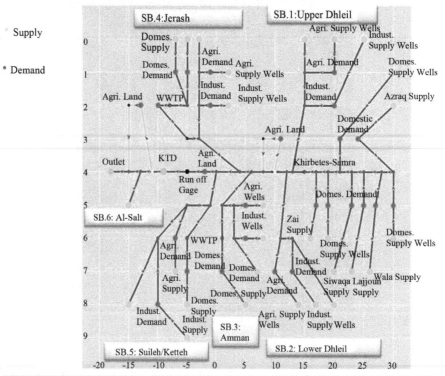

Figure 4. Topology Map of the Zarqa River Basin.

Swaqa, and Zai. The surface water originated from the basin as flood flow mixed with treated water are used partly for irrigation along the river and the rest are stored in King Talal Dam to be released for irrigation downstream and in the Jordan valley. **Table 2** and **Table 3** show the inter-basin transfer in MCM and the hydraulic load of the wastewater treatment plants in m³/sec.

Furthermore, for each functional node type of the WRM, all input and output data as time series are stored in the data base with its respective node for the hydrological year 2001/2002 to facilitate processing, results, and analysis. The application of WRM for ZRB has been presented in a previous paper [4] with complete description of its components that includes nodes and reaches. The network of ZRB is translated and conceptualized into a topology map of ninety five input-output nodes as shown in **Figure 4**.

Table 2. Supply from Inter-Basin Transfer (MCM) for the 2001/2002.

Supply from Inter-Basin Transfer	MCM
Azraq Domestic Supply	16.5
Wala Domestic Supply	11.70
Lajoun Domestic Supply	9.90
Swaqa Domestic Supply	9.0
Zai Domestic Supply	45.3
Total in MCM	92.4

Table 3. The hydraulic load of the wastewater treatment plants for 2001/2002.

Waste Water Treatment Plant (WWTP) Name	Hydraulic Capacity (Maximum Flow), m³/sec
Khirbet es-Samra WWTP-Located in SB 1	2.4
Abu Nssair WWTP-Located in SB 3	0.023
Baqa'a WWTP-Located in SB 5	0.16
Jerash WWTP-Located in SB 4	0.04

4. Simulation Results and Analysis

4.1. HEC-1 Model

The data base describes Zarqa Basin in monitoring time series of rainfall, runoff data, and the main components of the HEC-1 model *i.e.* the curve numbers. Referring to Shammout, 2003 [2], the average curve numbers for Zarqa River Basin under dry condition are 59, 78 under a normal condition and 89 under a wet condition. During vegetation period, the average curve numbers under dry condition are 52, 72 under a normal condition and 86 under a wet condition.

HEC-1 is run using the huge capability of WMS. The data used for running the model starts from 1/10/2001 to 30/5/2002. **Table 4** shows the computed sub-basin's area, slope, and the rainfall weighted average. It can be seen from **Table 4** that the calculated rainfall weighted average in MCM for Upper Dhleil, Lower Dhleil, Amman, Jerash, Suileh/ Ketteh, and Al-Salt sub-basins are 293.11, 32.98, 245.0, 61.49, 147.67, and 130.9 respectively, where the total rainfall is 911.15 MCM. The certainty of HEC-1 application will be deduced from the compatibility of the modeled flow to the actual flow data. For instance, the availability of daily runoff data allows the calibration of the modeled flow with known data series [2]. Therefore, the results of running HEC-1 model are achieved and show that the annual predicted (modeled) runoff volume in MCM is 36.4 from the total rainfall storms. In comparison with measured (actual) total runoff volume of the Runoff Gauge at Jerash Bridge which is 36.6 MCM as shown in **Table 5**, is compatible to the modeled runoff. Thus, the predicted runoff with a value of 36.4 MCM very well matches the figures of the Ministry of Water and Irrigation [13] and it excludes the effluent of Khirbet es-Samra WWTP.

4.2. Water Resources Model

WRM is run online using the huge capability of ESS server. **Table 6** shows the computed water resources in MCM of ZRB using WRM for each SBs.

It can be seen from **Table 6** that the calculated surface runoff (flood + base flow) in MCM via WRM for Upper Dhleil, Lower Dhleil, Amman, Jerash, Suileh/ Ketteh, and Al-Salt sub-basins are 6.53, 5.6, 26.13, 28, 28, and 2.8 respectively. The total volume contribution of surface runoff of the 6-sub-basins is 97.1 MCM including the effluent from the treatment plants *i.e.* (base flow + flood). Besides, the flood gauge station of Jerash Road Bridge is the only gauge in watershed that has complete runoff records and is considered as the control node.

Table 4. The sub-basin's area, slope, and the rainfall weighted average.

Parameter	Upper Dhleil SB 1	Lower Dhleil SB 2	Amman SB 3	Jerash SB 4	Suileh/Ketteh SB 5	Al-Salt SB 6	Total
Area Km^2	2120.49	221.92	658.82	503.34	323.03	291.52	4119.12
BS m/m	0.0277	0.0585	0.0966	0.1724	0.1990	0.2266	-
Rainfall Weighted Average MCM	293.11	32.98	245.0	61.49	147.67	130.9	911.15

Table 5. Predicted and measured total runoff volume.

MCM	Predicted Runoff (Modeled)	Measured Runoff of the Runoff Gauge at Jerash Bridge
Annual Surface Runoff Volume **Excluding** the Effluent of Khirbet es-Samra WWTP- using HEC-1 Model	36.4	36.6

Table 6. Water resources of Zarqa River Basin for each sub-basin of the 2001/2002.

Water Supply (MCM)	Upper Dhleil SB 1	Lower Dhleil SB 2	Amman SB 1	Jerash SB 4	Suileh/Ketteh SB 5	Al-Salt SB 6	Total
Domestic Wells	34.84	8.45	26.2	1.45	4.52	-	75.46
Industrial Wells	0.45	3.36	1.6	0.01	0.26	-	5.69
Agriculture Wells	42.01	2.04	2.96	2.3	3.21	-	52.52
Surface Runoff Contribution of Each SB **Including** the Effluent of WWTPs	6.53	5.6	26.13	28	28	2.8	97.1

Furthermore, the wastewater treatment plant at Khirbet es-Samra can also be considered as a control node for model certainty since the influent amounts as a daily record can be compared with wastewater discharge of the sewer systems. **Figure 5** shows the actual data and location of the control node at Jerash Bridge, and **Figure 6** shows the analysis of control nodes of Runoff Gauge at Jarash Bridge and Khirbet es-Samra WWTP.

Figure 6 shows the analysis of the control node where the average modeled flow is 2.97 m^3/s, which is compatible to the actual measured flow at the gauging station (Jarash Bridge) with a value of 2.96 m^3/s. Thus, the total inflow is 93.68 MCM (base flow and flood), which very well matches the figures of the Ministry of Water and Irrigation with a value of 93.32 MCM. Furthermore, it can be seen from **Figure 6** that the modeled wastewater treatment plant node at khirbet es-Samra is 63.53 MCM with a maximum flow of 2.43 m^3/sec, which matches the figures of the actual data (**Table 3**) with a value of 2.4 m^3/sec.

4.3. Comparison between HEC-1 Model and WRM

Comparing the results of WRM and HEC-1 models proved their simulation efficiency in predicting the flow of Zarqa River Basin. However, the philosophy of HEC-1 is a single storm event and is based on values of curve number, while WRM philosophy describes the water flow and availability, and demand and supply balance on a daily basis across the basin as shown in **Table 7**.

5. Conclusions

Assessing different modeling tools to produce comparative analysis is achievable and helps considerably the decision makers in managing watershed of scarce water resources. WMS proved its efficiency in preparing the initial models' requirements and is used to define watershed characteristics using a Digital Elevation Model (DEM). WMS is used in determining the flow direction and accumulation, stream network, basin outlet, and interior outlets, for creating the entire watershed and its sub-basins. The average rainfall and area/elevation can be derived for each sub-basin based on hydrologic sub-basin parameters. The results obtained from WMS are used as an input for models; HEC-1 Model, and the Water Resources Model (WRM).

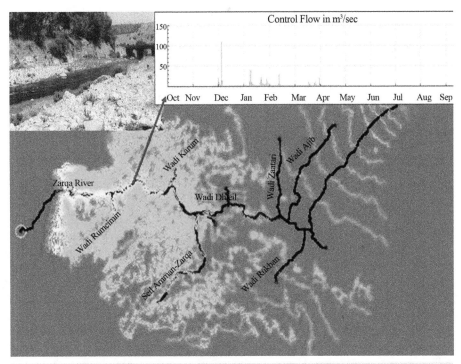

Figure 5. The actual flow data of the control node at Jerash Bridge.

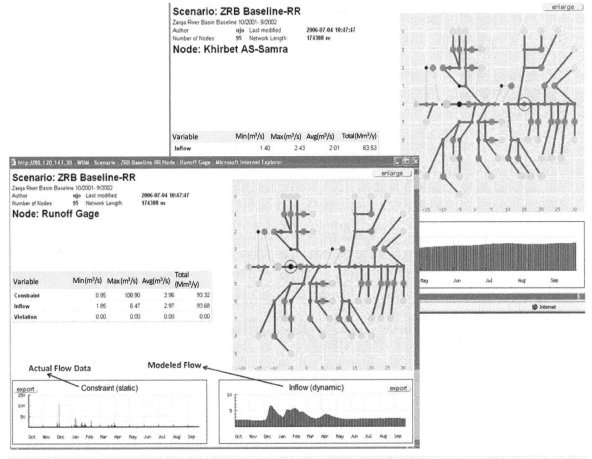

Figure 6. The analysis of the control nodes.

Table 7. Comparison between HEC-1 Model and WRM Model.

Models Comparison	HEC-1 Model	WRM
Philosophy	-Precipitation-Runoff Process. -Single Storm Event. -Based on CN.	WRM Describes the Water Flow, Storage and Availability, Demand and Supply Balance on a Daily Basis across a River Basin and Its Elements.
Predicted Annual Surface Runoff Volume in MCM.	36.4 Excluding the Effluent of Khirbet es-Samra WWTP	93.68 (Base flow and Flood)
Predicted Inflow of Khirbet es-Samra WWTP in m^3/sec.	-	2.43 (Maximum) 2.01 (Average)

The results of running HEC-1 model are achieved and show that the annual predicted (modeled) runoff volume in MCM is 36.4 from the total rainfall storms. In comparison with measured (actual) total runoff volume of the Runoff Gauge at Jerash Bridge which is 36.6 MCM, is compatible to the modeled runoff. Whereas, the results of running WRM show that the total inflow is 93.68 MCM (base flow and flood), which very well matches the figures of the Ministry of Water and Irrigation with a value of 93.32. Besides, the modeled wastewater treatment plant node at khirbet es-Samra is 63.53 MCM with maximum flow of 2.43 m^3/sec which matches the figures of the actual data with a value of 2.4 m^3/sec.

Comparing the results of WRM and HEC-1 models proved their simulation efficiency in predicting the flow of Zarqa River Basin. Nevertheless, the philosophy of HEC-1 is a single storm event and is based on values of curve number, while WRM philosophy describes the water flow and availability, and demand and supply balance on a daily basis across the basin. The models' predictions for the real flow definitely establish the modeling certainty and help water resources developers to incorporate different basin features for watershed representation, simulation, and management. Hence, the certainty of the results in modeling provides indirect ways of assessing the success of models' simulations.

Acknowledgements

The application on WRM was carried out with financial support from the European Commission under FP6 INCO-MED Program; contract INCO-CT-2004-509091.

References

[1]   Jasem, A.H., Shammout, M., AlRousan, D. and AlRaggad, M. (2011) The Fate of Disi Aquifer as Stratigic Groundwater Reserve for Shared Countries (Jordan and Saudi Arabia). *Journal of Water Resource and Protection*, **3**, 711-714. http://dx.doi.org/10.4236/jwarp.2011.310081

[2]   Shammout, M.W. (2003) Land Use Options for Surface Water Management in Zarqa River Basin Using Modeling Tools. Ph.D. Dissertation, The University of Jordan, Amman.

[3]   Shatanawi, M., Shammout, M.W. and Naber, S. (2008) Water Conflicts among Sectors and Environmental Uses in Jordan. *OPTIONS Méditerranéennes, Mediterranean Seminars, Water Culture and Water Conflict in the Mediterranean Area, Series A*, **83**, 159-172. http://ressources.ciheam.org/util/search/detail_numero.php?mot=384&langue=en

[4]   Shatanawi, M. and Shammout, M.W. (2011) Supply-Demand Modeling of Water Resources in Zarqa River Basin in Jordan. *International Journal of Applied Environmental Sciences (IJAES)*, **6**, 261-278. http://www.ripublication.com/Volume/ijaesv6n3.htm

[5]   DOS (2011) Statistical Year Book 2011. Department of Statistics, Amman. http://www.dos.gov.jo

[6]   Corps of Engineers (1998) HEC-1 Flood Hydrograph Package. User's Manual, Hydrologic Engineering Centre, US Army Corps of Engineers, Davis.

[7]   Chow, V.T., Maidment, D.R. and Mays, L.W. (1988) Applied Hydrology. International Edition, McGraw-Hill Book Company, New York.

[8]   http://www.ess.co.at/MANUALS/WATERWARE/wrmmodel.html

[9]   http://www.ess.co.at/WATERWARE

[10] http://www.ess.co.at/OPTIMA/

[11] http://www.ess.co.at/OPTIMA/cases.html

[12] Shammout, M.W., Shatanawi, M. and Naber, S. (2013) Participatory Optimization Scenario for Water Resources Management: A Case from Jordan. *Water Resources Management*, **27**, 1949-1962. http://dx.doi.org/10.1007/s11269-013-0264-9

[13] Ministry of Water and Irrigation (2002) Open Files. Amman.

4

Regeneration Responses in Partially-Harvested Riparian Management Zones in Northern Minnesota

Douglas N. Kastendick[1*], Brian J. Palik[1], Eric K. Zenner[2], Randy K. Kolka[1], Charles R. Blinn[3], Joshua J. Kragthorpe[1]

[1]USDA Forest Service, Northern Research Station, Grand Rapids, USA
[2]Penn State University, University Park, USA
[3]University of Minnesota, St. Paul, USA
Email: [*]dkastendick@fs.fed.us, bpalik@fs.fed.us, eric.zenner@psu.edu, rkolka@fs.fed.us, cblinn@umn.edu, jkragthorpe@fs.fed.us

Abstract

Trees serve important functions in riparian areas. Guidelines often suggest how riparian forests should be managed to sustain functions, including tree retention and increasing the component of conifers and later-successional species. While regeneration of early successional species is not discouraged, there is uncertainty about the ability to regenerate the latter along with more desirable species. We investigated the regeneration of species differing in successional status and growth forms under different amounts of residual basal area. The study was conducted in riparian sites in northern Minnesota USA. At each site, one portion of the riparian area was uncut, while a downstream area was harvested to 16 or 8 $m^2 \cdot ha^{-1}$. Woody vegetation was sampled before and five-years after harvesting and summarized as early, mid-, and late successional hardwoods, as well as conifers and shrubs. After five years, the density of early successional trees was lower at 16 $m^2 \cdot ha^{-1}$ compared to 8 $m^2 \cdot ha^{-1}$; densities in both treatments were lower than in clearcuts. Densities of mid- and late successional hardwoods and conifers did not increase in either treatment. The higher basal area treatment resulted in a lower density of shrubs, which might be important for establishing more desirable tree species, although this may require additional activities to promote establishment.

Keywords

Riparian Management, Overstory Retention, Regeneration, RMZ, Riparian Forest

---

[*]Corresponding author.

1. Introduction

Trees in riparian areas serve many important ecological functions, including shading of streams, bank stabilization and protection from erosion, interception and uptake of water and nutrients from the upland [1]-[4]. Additionally, riparian trees contribute energy in the form of particulate and dissolved organic matter to aquatic systems [5] and they are the future source of coarse woody debris in streams and lakes [6]. Riparian forests are important habitat for species that are dependent on forests in close proximity to water [7] [8].

Many organizations and agencies have guidelines on how riparian forests should be managed to sustain ecological functions [9] [10]. Typically, these guidelines stress retention of trees in the riparian management zone (RMZ) and provide recommendations about minimum basal areas to be retained during harvesting. Also, guidelines often recommend sustaining or enhancing the abundance of conifers and longer-lived, often later successional, hardwood species [9]. The greater longevity of conifers and later successional hardwoods compared to early successional species provides for greater continuity of riparian function and larger potential tree sizes, resulting in higher inputs of coarse woody debris into aquatic habitats in the long run.

While management guidelines generally do not discourage silvicultural activities in riparian areas, managers may nevertheless forego harvests in RMZs because of uncertainties about the ability to regenerate desirable tree species (*i.e.*, conifers, later successional species) and the ability to maintain riparian functions that are dependent on trees [11]. There have been only a few experimental studies that have examined regeneration responses to partial harvesting treatments of RMZs within the context of riparian management guidelines that aim to sustain tree-derived functions and increase the component of more desirable tree species [e.g. [12] [13]].

Here we experimentally investigated the regeneration response to partial harvesting treatments in riparian areas that prescribed different levels of residual basal area within fixed-width RMZs. Specifically, we investigated how species of differing successional status (early, mid-, late) and growth forms (conifers, hardwoods, shrubs) regenerated under residual basal areas of 16 $m^2 \cdot ha^{-1}$ and 8 $m^2 \cdot ha^{-1}$. This experiment was conducted in mature mixed-species riparian forests dominated by early successional species, including trembling aspen (*Populus tremuloides*) and paper birch (*Betula papyrifera*) and a smaller component of conifers and later successional hardwoods. Our applied goal was to provide managers with information about approaches for managing similar riparian areas in ways that might reduce the density of early successional species and increase the abundance of later successional and conifer species.

2. Materials and Methods

2.1. Study Area

This study was conducted in eight forested riparian areas in the Laurentian Mixed Forest Province of Minnesota. The province is as a broad ecotone between the eastern deciduous forest and boreal forest biomes. It has a temperate climate with mean annual temperatures between 1.1°C and 3.9°C, and average annual precipitation between 56 and 81 cm [14]. Soils originated from Pleistocene till [15] and include well drained loamy sands that are shallow to bedrock in the uplands and sandy loams in lower landscape positions.

The eight study sites were selected in 2003 to meet the following criteria: 1) riparian forests were located adjacent to perennial streams that were less than 6 m in width; 2) a minimum contiguous forested area of 6.5 ha with a minimum of 183 m of stream frontage; 3) forests were mature and well-stocked in both the riparian zones and adjacent uplands.

Prior to treatment, dominant tree species included paper birch, trembling aspen, balsam fir (*Abies balsamea*), black ash (*Fraxinus nigra*), sugar maple (Acer saccharum), red maple (*Acer rubrum*), and basswood (*Tilia americana*), with lesser amounts of northern red oak (*Quercus rubra*), bur oak (*Quercus macrocarpa*), green ash (*Fraxinus pennsylvanica*), white spruce (*Picea glauca*), black spruce (*Picea mariana*), big-tooth aspen (*Populus grandidentata*), balsam poplar (*Populus balsamifera*), yellow birch (*Betula alleghtniensis*), silver maple (*Acer saccharinum*), ironwood (*Ostrya virginiana*), and northern white cedar (*Thuja occidentalis*). Collectively, early successional species (paper birch, trembling aspen, big-tooth aspen, and balsam poplar) comprised around 52% of the total basal area.

2.2. Experimental Design and Treatments

At each study site (sites treated as blocks for statistical analysis), two 3.2 ha treatment stands were delineated on one side of the stream, with the two stands separated by at least 61 m of unharvested forest (**Figure 1**). Within each treatment stand, a 0.8 ha riparian management zone (RMZ) was delineated along the length of the stream (183 m) that extended 46 m towards the upland, a width that corresponds to recommended RMZ widths for even-age management in the state of Minnesota [16]. The remainder of the treatment unit (2.4 ha) was outside of the RMZ and considered to be upland forest.

In each block, the following treatments were assigned to the two experimental units: 1) upland clearcut-RMZ uncut (RMZC) and 2) upland clearcut-RMZ partially harvested (RMZH) to a residual basal area of 16 $m^2 \cdot ha^{-1}$ (33% reduction) or 3) upland clearcut-RMZ partially harvested (RMZL) to a residual basal area of 8 $m^2 \cdot ha^{-1}$ (66% reduction). Only one of the two RMZ harvest treatments was assigned to each block due to space constraints; the result was an incomplete block design (see Statistical Analysis). To lessen confounding impacts of harvesting on streams, the RMZC treatment was always established upstream of the RMZH or RMZL treatments. The tree retention levels tested in this study fell within the range of residual basal area values recommended for RMZs in Minnesota [16]. Marking of residual trees followed riparian guidelines for Minnesota by reserving, where possible, longer-lived species, conifers, and hard mast-producing species.

Timber harvesting operations were conducted by experienced operators on frozen ground when sufficient snow had accumulated during the winter of 2003-2004 using conventional harvesting equipment (*i.e.*, feller-buncher and grapple skidder).

2.3. Vegetation Sampling

In each stand, five transects were established running perpendicular to the average stream meander, originating at the stream bankfull edge and terminating in the upland (**Figure 1**). Five vegetation measurement plots were established on each transect; four plots (each 4.6 m wide by 7.6 m long = 34.8 m^2) located in the RMZ and one plot located in the upland area outside of the RMZ (**Figure 1**). The first RMZ plot was established at the stream bankfull edge. The distance between the remaining RMZ plots varied because plots were constrained to be centered within major geomorphic features (*i.e.*, floodplain, terrace, hillslope), but generally the second RMZ plot was 9.1 m. from the stream, while the third plot was 22.9 m. from the stream. The fourth RMZ plot was always located 3 m. inside of the RMZ-upland boundary. The upland plot was always established 22.9 m. outside of the RMZ-upland boundary. Each treatment unit contained a total of 25 plots (5 transects × 5 plots).

Woody vegetation was sampled using a nested plot design. Species and diameter at 1.4 m (breast height = dbh) of trees (dbh ≥ 12.7 cm) and saplings (2.5 ≤ dbh < 12.7 cm) were sampled in the 4.6 m × 7.6 m plots. Woody stems (shrub species and advance tree regeneration) with dbh < 2.5 cm but ≥0.8 m tall were sampled in two 0.6 m × 4.6 m (2.8 m^2) plots nested within ends of the larger overstory plot. Small woody stems (<0.8 m tall) were tallied in six 0.6 m × 0.6 m (0.36 m^2) regeneration plots nested within each of the 2.8 m^2 plots. Vegetation was measured in the summer before harvesting and again five years after the harvest, after the majority of the growing season had elapsed, but prior to leaf senescence.

2.4. Statistical Analysis

Due to the variation in composition among the eight study blocks and to facilitate statistical analyses, we com-

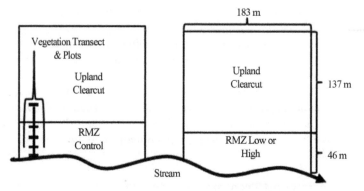

Figure 1. Study site layout illustrating treatment unit dimensions and vegetation sampling plot layout.

bined species into broader groups that reflected successional status and growth form. Groups included early successional hardwoods (paper birch, trembling aspen, big-tooth aspen, balsam poplar), mid-successional hard-woods (black ash, northern red oak, bur oak, green ash, silver maple, yellow birch, and American elm), late successional hardwoods (sugar maple, red maple, basswood, ironwood), and conifers (balsam fir, white spruce, black spruce, northern white cedar). Woody shrubs included mountain maple (*Acer spicatum*), alder (*Alnus* spp.), serviceberry (*Amelanchier* spp.), dogwood (*Cornus alternifolia, Cornus sericea*), hazel (*Corylus americana, Corylus cornuta*), honeysuckle (*Lonicera* spp.), cherry (*Prunus pensylvanica, Prunus virginiana*), and willow (*Salix* spp.). We summarized changes (between pre-harvest and 5th year post-harvest) in the proportional contri-bution to total density (in the sapling and regeneration layers) of each of these broad species groups (early, mid-, late successional, conifers, shrubs). For example, early successional hardwoods might have contributed 20% to total sapling density before harvest, but 60% after harvest, for a change of +40%.

Differences among treatments in basal area and stem density were assessed using an incomplete block mixed-model ANOVA with eight replicates for the control treatment (RMZC) and four replicates each for the riparian harvest treatments (RMZH, RMZL). Block was the random effect and basal area was the fixed effect. Pre- and post-harvest responses were analyzed separately. Some variables, including pre-harvest late success-sional regeneration, 5th year early successional saplings, mid-successional trees, mid-successional saplings, late successional saplings, and regeneration were square-root transformed to meet the assumption of normality and homogeneity of variances. An alpha level of ≤ 0.05 was considered statistically significant. Statistically signifi-cant models were further investigated using a set of orthogonal contrasts that compared 1) the RMZCs to the pooled partial harvesting treatments and 2) the RMZHs to the RMZLs. All statistical analyses were conducted using SAS 8.02 [17]. For comparative purposes, we included data from the adjacent upland clearcuts in the re-sults, but we did not include these data in the statistical analyses because the upland clearcut treatment was con-founded with location in our design.

3. Results

3.1. Tree Layer

Mean pre-harvest basal area was 23.7 m^2·ha^{-1} and did not differ statistically among the (future) RMZ treatments (p > 0.05; **Table 1**). Densities of early, mid- and late successional hardwood species, as well as conifers and to-tal density, were not significantly different among treatments before harvest. Five years after harvest, residual basal areas reflected the harvest intensity gradient from the RMZC to the RMZL treatments (**Table 1**). Addi-tionally, there was a statistically significant difference among treatments in total tree density (p = 0.025), with higher densities in the RMZCs than in the pooled harvest treatments (p = 0.01), but no significant differences between the two RMZ treatments (p > 0.05). There were trends in densities among treatments for all species groups that largely paralleled the gradient in basal area reduction (**Table 1**); however, due to high variation, these densities were not statistically different among the treatments. As expected, tree densities for all groups of trees and total density were substantially lower in the adjacent clearcuts than in the RMZ treatments.

3.2. Sapling Layer

Before harvest, total sapling density, as well as densities of hardwoods, conifers, and woody shrubs (that were large enough to be classified as saplings) did not differ statistically among (future) treatments (**Figure 2(a)**). Five years after harvest, total density of saplings did not differ among treatments, but there were differences among treatments in early successional hardwoods (p = 0.03; **Figure 2(b)**), with higher density in the RMZL compared to the RMZH treatment (p = 0.012). Moreover, early successional hardwood densities were higher in the clearcut uplands than either of the RMZ harvest treatments. The change in proportional contribution to total sapling density of early successional hardwoods after five years was small (< ±5%) in the RMZC and RMZH treatments, but was high and positive in the RMZL treatment (+13%) and the clearcut uplands (+56%), reflect-ing the substantial increases in sapling density of this group in these treatments (**Table 2**).

Densities of mid- and late successional hardwoods declined from the control to the RMZL treatment (**Figure 2(b)**), but the differences among treatments were not statistically significant. Proportional contribution to total sapling density of these groups did not change appreciably (≤ ±5%) by five years after harvest (**Table 2**).

Conifer densities were not statistically different among RMZ treatments (**Figure 2(b)**) and changes in propor-tional contribution to total sapling density were low (< ±5%). In contrast, the proportional contribution of

Table 1. Tree layer[a] characteristics in riparian management zone treatments.

	Pre-Harvest			Post-Harvest (5 years)			Upland Clearcut	
	RMZC[b]	RMZH[c]	RMZL[d]	RMZC	RMZH	RMZL	Pre	Post
Basal Area (m²·ha⁻¹)	23.9 (1.8)[e]	25.7 (3.9)	21.6 (4.6)	23.7 (2.1)	15.6 (2.5)	8.0 (1.6)	23.9 (2.5)	1.1 (0.5)
Total Density (stems·ha⁻¹)	1486 (114)	1463 (161)	1366 (116)	1431 (163)	877 (156)	576 (183)	1463 (109)	185 (69)
Early Successional[f] Density (stems·ha⁻¹)	613 (119)	675 (217)	682 (185)	536 (126)	284 (124)	213 (116)	771 (143)	62 (35)
Mid-Successional[g] Density (stems·ha⁻¹)	284 (163)	274 (252)	106 (72)	274 (158)	222 (198)	62 (22)	96 (49)	35 (27)
Late Successional[h] Density (stems·ha⁻²)	195 (91)	195 (104)	222 (222)	213 (101)	161 (82)	62 (42)	292 (148)	0 (0)
Conifers[i] (stems·ha⁻¹)	395 (126)	319 (156)	356 (193)	408 (158)	213 (104)	240 (124)	301 (96)	89 (47)

[a]Diameter at 1.4 m ≥ 12.7 cm; [b]RMZC = uncut (control) RMZ; [c]RMZH = RMZ cut to 16 m²·ha⁻¹; [d]RMZL = RMZ cut to 8 m²·ha⁻¹; [e]Means ± 1 standard error; [f]Early successional includes paper birch, trembling aspen, big-tooth aspen, and balsam poplar; [g]Mid-successional includes black ash, green ash, northern red oak, bur oak, silver maple, yellow birch, and American elm; [h]Late successional includes sugar maple, red maple, basswood, and ironwood; [i]Conifers includes balsam fir, white spruce, black spruce, and northern white cedar.

conifers to total sapling density declined by nearly 19% in the clearcut (**Table 2**). Woody shrub densities were reduced by harvest (**Figure 2(b)**), although they did not differ statistically among RMZ treatments. Densities were lower in the upland clearcuts compared to the RMZ treatments. Proportional contribution of shrubs to total sapling density declined in all treatments. The decline was small (−3.9%) in the RMZC treatment (**Table 2**), but larger in the other RMZ treatments (−11.3% in RMZH; −15.9% in RMZL) and the upland clearcut (−29.7%).

3.3. Regeneration Layer

Densities of hardwood groups were low before harvest and did not differ statistically among (future) treatments (**Figure 3(a)**). Densities of conifers and woody shrubs were higher than those of hardwoods, but were variable and their densities, as well as total regeneration density, did not differ statistically among (future) treatments. Five years after harvest, total density in the regeneration layer differed statistically among treatments (p = 0.04), with higher densities in the pooled harvest treatments compared to the RMZC treatment (p = 0.03), but no statistical difference between the two partially harvested RMZ treatments. Densities of early successional hardwoods after harvest differed statistically among treatments (p < 0.001; **Figure 3(b)**), with higher densities in the pooled harvest treatments compared to the RMZC treatment (p = 0.002) and higher densities in the RMZL compared to the RMZH treatment (p = 0.03). Densities in the clearcut uplands were substantial higher than in the partially harvested RMZ treatments. Change in proportional contribution of early successional hardwoods to total regeneration density was positive for all treatments (**Table 2**), but not particularly high (< +5.4%), indicating that the relative densities of this group were not substantially changed by year five, mostly because sucker stems had grown into the sapling layer by this time.

Densities of mid- and late-successional hardwoods did not differ statistically among treatment five years after harvest and were similar to pre-harvest levels, with small changes in proportional contribution to total density (< ±5.3%), with the exception of late successional hardwoods in the RMZH treatment, which were substantially reduced (−11.9%). Conifer densities were reduced from pre-harvest levels in all treatments, including a 24% - 46% reduction in proportional contribution to total density (**Table 2**), but densities did not differ significantly among treatments.

Woody shrub densities differed among treatments five years after harvest (**Figure 3(b)**), with densities in the pooled harvest treatments statistically higher than in the RMZC treatment (p = 0.03) and densities in the RMZL treatment higher than in the RMZH treatment (p = 0.04). A commensurate large increase (21.1% to 40.6%) in the proportional contribution of shrubs to total regeneration density was observed in all treatments (**Table 2**).

4. Discussion

An important question for forest managers is how to balance extracting timber from riparian areas with sustaining riparian functions that depend on the regeneration of desirable tree species and species groups (e.g., con-

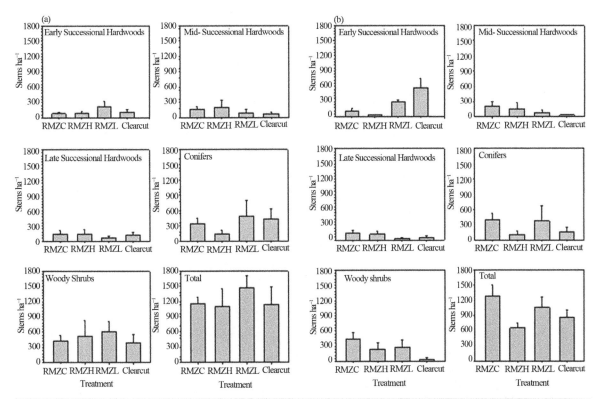

Figure 2. Sapling (2.5 < dbh < 12.5 cm) densities of species groups before harvest (a) and five years after harvest (b) See text for explanation of treatments and species groups.

Table 2. Change (from pre-harvest to five years after harvest) in proportional contribution to total density of different successional classes and growth forms in the different treatments.

Treatment[a]	Sapling Layer					Regeneration Layer				
	Early[b]	Mid-[c]	Late[d]	Conifers[e]	Shrubs[f]	Early	Mid-	Late	Conifers	Shrubs
RMZC	1.9	2.4	−2.0	1.6	−3.9	1.4	5.3	3.9	−31.9	21.1
RMZH	−3.0	4.8	4.6	4.8	−11.3	3.7	1.9	−11.9	−23.9	30.1
RMZL	13.2	1.3	−1.1	2.5	−15.9	3.7	1.7	0.5	−45.9	40.0
Upland										
Clearcut	56.1	−2.6	−5.2	−18.6	−29.7	5.4	−0.9	−5.1	−40.1	40.6

[a]RMZC = uncut (control) RMZ; RMZH = RMZ cut to 16 m^2·ha^{-1}; RMZL = RMZ cut to 8 m^2·ha^{-1}. [b]Early successional includes paper birch, trembling aspen, big-tooth aspen, and balsam poplar. [c]Mid-successional includes black ash, northern red oak, bur oak, green ash, silver maple, yellow birch, and American elm. [d]Late successional includes sugar maple, red maple, basswood, and ironwood. [e]Conifers includes balsam fir, white spruce, black spruce, and northern white cedar. [f]Shrubs include the woody species mountain maple, speckled alder, serviceberry, dogwood, honeysuckle, cherry, willow, and hazel species.

ifers, later successional species). This is particularly important in the water-rich Great Lakes regional landscape where a high percentage of the forest is riparian. For example, one estimate indicates that 37% of commercial forests in Minnesota are within 57 m of surface water [18], a distance that places the bulk of this forest within Minnesota's recommended RMZ width guideline of 46 m [16]. Consequently, a management objective for a mixed riparian forest like the one we examined will likely include managing for timber but in ways that: 1) reduce the density of early successional species, which in the Lake States often includes aspen and birch; and 2) regenerating and enhancing the proportion of longer lived, late-successional hardwoods and conifers.

The partial harvest treatments in the RMZs we examined were successful at meeting the first part of this ob-

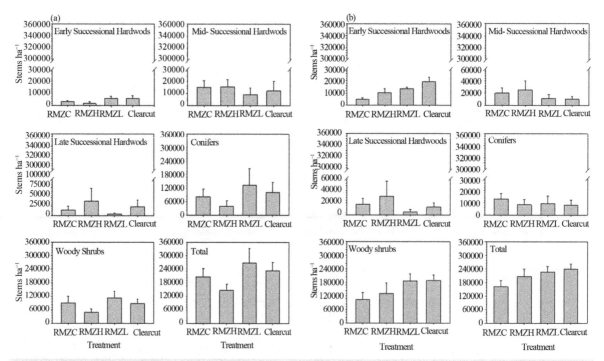

Figure 3. Regeneration (dbh < 2.5 cm) densities of species groups before harvest (a) and five years after harvest (b) See text for explanation of treatments and species groups.

jective. While early results (1 and 3 years after harvest) indicated that aspen sucker densities in the RMZs were within the range of acceptable stocking [13], after five years, this stocking had declined to levels that were less than one-third the stocking densities for aspen in similar aged clearcut stands [19] [20] and much less than the densities in the adjacent clearcuts we examined, especially in the RMZH treatment. While the density of early successional taxa did increase in parallel with the decreasing basal area of residual trees (from RMZC to RMZL), and the proportional composition of total sapling density for aspen and birch increased by 13.2% in the RMZL treatment, these relatively low basal areas still apparently checked sucker and seedling establishment or survival, relative to clearcuts, similar to findings from other studies [21]-[23].

The second part of the management objective, that of enhancing the abundance of later successional hardwoods and conifers, was largely not accomplished. Despite reserving mid- and later successional hardwood species and conifers as future seed sources in the overstory, these species were not able to capitalize (at least in the first 5 years) on the reduced density of early successional hardwoods. In fact, the proportional contribution to total regeneration density of conifers declined substantially in all RMZ treatments, including the controls, with a particularly strong decline in the RMZL treatment. We hypothesize that increases in solar radiation and temperature, which may have affected conifers even in the controls through edge effects [24] [25], may have caused moisture stress and mortality in spruce and fir seedlings [26] [27].

The increase in proportional contribution of woody shrubs, mainly beaked hazel and mountain maple, to total density in the regeneration layer of all RMZ treatments, and particularly in the RMZL treatment, may have further contributed to the limited regeneration success of trees. Resource preemption by a woody understory layer [28]-[30], and beaked hazel in particular [31], is recognized as a deterrent to the regeneration of trees. Thus the response of shrubs to harvesting in this study may have restricted the establishment of new trees and/or the growth of seedlings into the sapling size class. This inhibition appears to have been somewhat less in the RMZH compared to RMZL treatment.

While our treatments did not result in significant increases in densities of trees species deemed especially desirable in riparian settings (mid- and later successional hardwoods, conifers), our results do point to some practical guidance to managers working in such settings. First, leaving a residual basal area of 16 m$^2 \cdot$ha^{-1}, as we did in our RMZH treatment, was effective at limiting the regeneration density of early successional species (mostly aspen) to a greater degree than 8 m$^2 \cdot$ha^{-1} in the RMZL treatment. Ultimately this reduction should be important for creating greater opportunities for establishment and growth of the other species groups. Secondly, although the density of sapling-sized woody shrubs declined following treatment, woody shrub density increased substantially in the regeneration layer. The increased abundance of these keen competitors may be an obstacle to the establishment, survival, and growth of more desirable tree species in riparian areas. This combination of responses—suppression of early successional species and expansion of woody shrubs, suggests the need for additional activities to promote regeneration of mid- and later successional hardwoods and conifers. In particular, competition control to reduce shrub densities [12] [13], along with seeding or planting of desired species, may be needed to ensure regeneration success.

Acknowledgements

The authors thank Dwight Streblow, Adam Sutherland, and Jeff Killmer for logistical and field support, the USDA Forest Service Northern Research Station, the University of Minnesota, Department of Forest Resources, and the Minnesota Environment and Natural Resources Trust Fund for financial support.

References

[1] Kreutzweiser, D.P. and Capell, S.S. (2001) Fine Sediment Deposition in Streams after Selective Forest Harvesting without Riparian Buffers. *Canadian Journal of Forest Research*, **31**, 2134-2142. http://dx.doi.org/10.1139/x01-155

[2] Broadmeadow, S. and Nisbet, T.R. (2004) The Effects of Riparian Forest Management on the Freshwater Environment: A Literature Review of Best Management Practices. *Hydrology and Earth System Science*, **8**, 286-305. http://dx.doi.org/10.5194/hess-8-286-2004

[3] Hazlett, P.W., Gordon, A.M., Voroney, R.P. and Sibley, P.K. (2007) Impact of Harvesting and Logging Slash on Nitrogen and Carbon Dynamics in Soils from Upland Spruce Forests in Northeastern Ontario. *Soil Biology and Biochemistry*, **39**, 43-57. http://dx.doi.org/10.1016/j.soilbio.2006.06.008

[4] Kastendick, D.N., Zenner, E.K., Palik, B.J., Kolka, R.K. and Blinn, C.R. (2012) Effects of Harvesting on Nitrogen and Phosphorus Availability in Riparian Management Zone Soils in Minnesota, USA. *Canadian Journal of Forest Research*, **42**, 1784-1791. http://dx.doi.org/10.1139/x2012-127

[5] Naiman, R.J., Bilby, R.E. and Bisson, P.A. (2000) Riparian Ecology and Management in Pacific Coastal Rain Forest. *BioScience*, **50**, 996-1011. http://dx.doi.org/10.1641/0006-3568(2000)050[0996:REAMIT]2.0.CO;2

[6] Palik, B.J., Golladay, S.W., Goebel, P.C. and Taylor, B.W. (1998) Geomorphic Variation in Riparian Tree Mortality and Stream Coarse Woody Debris Recruitment from Record Flooding in a Coastal Plain Stream. *Ecoscience*, **5**, 551-560.

[7] Hannon, S.J., Paszkowski, C.A., Boutin, S., DeGroot, J., Macdonald, S.E., Wheatley, M. and Eaton, B.R. (2002) Abundance and Species Composition of Amphibians, Small Mammals, and Songbirds in Riparian Forest Buffer Strips of Varying Widths in the Boreal Mixedwood of Alberta. *Canadian Journal of Forest Research*, **32**, 1784-1800. http://dx.doi.org/10.1139/x02-092

[8] Goebel, C.P., Palik, B.J. and Pregitzer, K.S. (2003) Plant Diversity Contributions of Riparian Areas in Watersheds of the Northern Lake States, USA. *Ecological Applications*, **13**, 1595-1609. http://dx.doi.org/10.1890/01-5314

[9] Blinn, C.R. and Kilgore, M.A. (2001) Riparian Management Practices: A Summary of State Guidelines. *Journal of Forestry*, **99**, 11-17.

[10] Wisconsin Department of Natural Resources (2011) Chapter 5: Riparian Areas and Wetlands. Pub-FR-226, Madison.

[11] Palik, B.J., Zasada, J.C. and Hedman, C.W. (2000) Ecological Principles for Riparian Silviculture. In: Verry, E.S., Hornbeck, J.W. and Dolloff, C.A., Eds., *Riparian Management in Forests of the Continental Eastern United States*, Lewis Publishers, Boca Raton, 233-254.

[12] Palik, B., Martin, M., Zenner, E., Blinn, C. and Kolka, R. (2012) Overstory and Regeneration Dynamics in Riparian Management Zones of Northern Minnesota Forested Watersheds. *Forest Ecology and Management*, **271**, 1-9. http://dx.doi.org/10.1016/j.foreco.2012.01.045

[13] Zenner, E.K., Olszewski, S.L., Palik, B.J., Kastendick, D.N., Peck, J.E. and Blinn, C.R. (2012) Riparian Vegetation Response to Gradients in Residual Basal Area with Harvesting Treatment and Distance to Stream. *Forest Ecology and Management*, **283**, 66-76. http://dx.doi.org/10.1016/j.foreco.2012.07.010

[14] Minnesota Department of Natural Resources (2003) Field Guide to the Native Plant Communities of Minnesota: The Laurentian Mixed Forest Province. Ecological Land Classification Program, Minnesota County Biological Survey, and Natural Heritage and Nongame Research Program, St. Paul.

[15] Keys Jr., J.E., Carpenter, C.A., Hooks, S.L., Koeneg, F.G., McNab, W.H., Russell, W.E. and Smith, M.L. (1995) Ecological Units of the Eastern United States—First Approximation. *USDA Forest Service Technical Publication*, R8-TP

21.

[16] Minnesota Forest Resources Council (1999) Sustaining Minnesota Forest Resources: Voluntary Site-Level Management Guidelines for Landowners, Loggers and Resource Managers. Minnesota Forest Resources Council, St. Paul.

[17] SAS (Statistical Analysis Systems Institute) (2002) SAS Ver. 8.02. SAS Institute, Cary.

[18] Hanowski, J., Danz, N., Lind, J., Niemi, G. and Wolter, P. (2001) Wildlife Species: Responses to Forest Harvesting and Management in Riparian Stands and Landscapes. Final Report to the Minnesota Forest Resources Council, St. Paul.

[19] Perala, D.A. (1979) Regeneration and Productivity of Aspen Grown on Repeated Short Rotations. USDA Forest Service Research Paper NC-176.

[20] Stone, D.M. and Elioff, J.D. (1998) Soil Properties and Aspen Development Five Years after Compaction and Forest Floor Removal. *Canadian Journal of Soil Science*, **78**, 51-58. http://dx.doi.org/10.4141/S97-026

[21] Huffman, R.D., Fajvan, M.A. and Wood, P.B. (1999) Effects of Residual Overstory on Aspen Development in Minnesota. *Canadian Journal of Forest Research*, **29**, 284-289. http://dx.doi.org/10.1139/x98-202

[22] Stone, D.M., Elioff, J.D., Potter, D.V., Peterson, D.B. and Wagner, R. (2001) Restoration of Aspen-Dominated Ecosystems in the Lake States. In: Shepperd, W.D., Ed., *Sustaining Aspen in Western Landscapes*, USDA Forest Service General Technical Report RMRS-P-18, USDA, Forest Service, Rocky Mountain Research Station, 137-143.

[23] Palik, B.J., Cease, C. and Egeland, L. (2003) Aspen Regeneration in Riparian Management Zones in Northern Minnesota: Effects of Residual Overstory and Harvest Method. *Northern Journal of Applied Forestry*, **20**, 79-84.

[24] Young, A. and Mitchell, N. (1994) Microclimate and Vegetation Edge Effects in a Fragmented Podocarp-Broadleaf Forest in New Zealand. *Biological Conservation*, **67**, 63-72. http://dx.doi.org/10.1016/0006-3207(94)90010-8

[25] Chen, J., Franklin, J.F. and Spies, T.A. (1995) Growing-Season Microclimatic Gradients from Clearcut Edges into Old-Growth Douglas-Fir Forests. *Ecological Applications*, **5**, 74-86. http://dx.doi.org/10.2307/1942053

[26] Thomas, P.A. and Wein, R.W. (1985) Water Availability and the Comparative Emergence of Four Conifer Species. *Canadian Journal of Botany*, **63**, 1740-1746.

[27] Burns, R.M. and Honkala, B.H. (1990) Silvics of North America: 1. Conifers; 2. Hardwoods. US Department of Agriculture, Forest Service, Agriculture Handbook 654, Washington DC.

[28] Lorimer, C.G. Chapman, J.W. and Lambert, W.D. (1994) Tall Understory Vegetation as a Factor in the Poor Development of Oak Seedlings beneath Mature Stands. *Journal of Ecology*, **82**, 227-237. http://dx.doi.org/10.2307/2261291

[29] Dovčiak, M., Frelich, P.B. and Reich, L.E. (2003) Seed Rain, Safe Sites, Competing Vegetation and Soil Resources Spatially Structure White Pine Regeneration and Recruitment. *Canadian Journal of Forest Research*, **33**, 1892-1904. http://dx.doi.org/10.1139/x03-115

[30] Weyenberg, S., Frelich, P.B. and Reich, L.E. (2004) Logging Versus Fire: How Does Disturbance Type Influence the Abundance of *Pinus strobus* Regeneration. *Silva Fennica*, **38**, 79-194.

[31] Kuuseoks, E., Dong, J. and Reed, D. (2001) Shrub Age Structure in Northern Minnesota Aspen Stands. *Forest Ecology and Management*, **149**, 265-274. http://dx.doi.org/10.1016/S0378-1127(00)00559-4

5

A Modified Pareto Dominance Based Real-Coded Genetic Algorithm for Groundwater Management Model

Li Fu

Shanghai Guanglian Construction Development Co. Ltd., Shanghai, China
Email: li_fu020@163.com

Abstract

This study proposes a groundwater management model in which the solution is performed through a combined simulation-optimization model. In the proposed model, a modular three-dimensional finite difference groundwater flow model, MODFLOW is used as simulation model. This model is then integrated with an optimization model, in which a modified Pareto dominance based Real-Coded Genetic Algorithm (mPRCGA) is adopted. The performance of the proposed mPRCGA based management model is tested on a hypothetical numerical example. The results indicate that the proposed mPRCGA based management model is an effective way to obtain good optimum management strategy and may be used to solve other type of groundwater simulation-optimization problems.

Keywords

Groundwater, Groundwater Management Model, Simulation-Optimization, Pareto Dominance, Genetic Algorithm

1. Introduction

Groundwater is a vital resource throughout the world. Nowadays, with increasing population and living standards, there is a growing need for the utilization of groundwater resources. Unfortunately, the quantity and quality of groundwater resources continues to decrease due to population growth, unplanned urbanization, industrialization, and agricultural activities. Therefore, sustainable management strategies need to be developed for the optimal management of groundwater resources [1]-[3].

Groundwater management models are widely used to determine the optimum management strategy by integrating optimization models with simulation models, which predict the groundwater system response [3]-[5].

Many researchers have adopted non-heuristic optimization approaches in conjunction with groundwater simulation models to solve groundwater management problems [6]-[9]. Typical problems in groundwater management problems are to maximize the total pumping or to minimize the total cost of capital, well drilling/installing and operating at a fixed demand [10]. But these optimization approaches may be not effective for problems that contain several local minima and for problems where the decision space is highly discontinuous [1] [2] [11].

Groundwater management problems are commonly nonlinear and non-convex mathematical programming problems [11]. In the last decades, many heuristic optimization approaches, based on the rules of the natural processes, have been proposed and utilized to deal with the groundwater management problems. Among these heuristic optimization approaches, the mostly widely used heuristic optimization approach is genetic algorithm (GA), which based upon the mechanism of biological evolutionary process.

Many studies deal with groundwater management problems using genetic algorithms. Mckinney and Lin (1994) integrated GA based optimization model with a groundwater simulation model programming to solve three management problems (maximum pumping problem, minimum cost pumping problem, and pump-and-treat design problem) [5]. Cieniawski *et al.* (1995) applied GA to optimize the groundwater monitoring network under uncertainty [12]. Wang and Zheng (1998) combined GA and SA (Simulated Annealing algorithm) based optimization model with MODFLOW model for maximization of pumping and minimization of the cost [10]. Wu *et al.* (1999) developed a GA based SA penalty function approach (GASAPF) to solve a groundwater management model [13]. Mahinthakumar and Sayeed (2005) solved a contaminant source identification problem by hybrid GAs that combine GA with different local search methods [14]. Bhattacharjya and Datta (2009) linked ANN (Artificial Neural Network) model with GA-based optimization model to solve multiple objective saltwater management problems [15]. The studies summarized above indicate that GA and GA-based approaches are good choices to solve groundwater management problems.

But similar to other heuristic optimization approaches, GAs are also unconstrained search technology and lack a clear mechanism for constraint handling [16]. Thus, their performance is blocked when dealing with nonlinear COPs (Constrained Optimization Problems) [17]. Groundwater management problems are usually nonlinear COPs. An appropriate constraint handling technique may increase the efficiency and effectiveness of GA and GA-based approaches for solving groundwater management problems.

In trying to solve COPs using GA or other optimization methods, penalty function methods have been the most popular approach [13] [18]-[20], because of their simplicity and ease of implementation. However, their performance is not always satisfactory, and the most difficult aspect of the penalty function method is to find appropriate penalty parameters needed to guide the search towards the constrained optimum [21] [22].

Thus, many researchers have developed sophisticated penalty functions or proposed other various constraint handling techniques over the past decade. Relevant methods proposed for constraint handling for heuristic optimization approaches can be categorized into: 1) penalty function methods; 2) methods based on preserving feasibility of solutions; 3) methods which make a clear distinction between feasible and infeasible solutions; and 4) hybrid methods [17] [23] [24].

Among these constraint handling techniques, methods based on multi-objective concepts have attracted increasing attention. Deb (2000) introduced a constraint handling method that requires no penalty parameters, this method used the following criteria: 1) any feasible solution is preferred to any infeasible solution; 2) between two feasible solutions, the one with better objective function value is preferred; and 3) between two infeasible solutions, the one with smaller degree of constraint violation is preferred [22]. Zhou *et al.* (2003) addressed on transforming single objective optimization problem to bi-objective optimization problem, with the first objective to optimize the original objective function, and the second to minimize the degree of constraint violation [25]. Mezura-Montes and Coello (2005) presented a simple multimembered evolution strategy to solve nonlinear optimization problems, and this approach also does not require the use of a penalty function [26]. To sum up, the main advantage of methods based on multi-objective concepts is avoiding the fine-tuning of parameters of penalty function.

However, it is worth noting that the newly-defined multi-objective problem (MOP), which is transformed from single objective COP, is in nature different from the customary MOP. That is, the philosophy of customary MOP is to obtain a final population with a diversity of non-dominated individuals, whereas the newly-defined MOP would retrogress to a single objective optimization problem within the feasible region [16].

In this study, methods based on multi-objective concepts are utilized to handle the constraints in groundwater management models. We firstly adopt multi-objective concept to transform single objective COPs to bi-objective optimization problems. Next, Pareto dominance is introduced for comparison of vectors and then individual's Pareto intensity number is used to substitute for fitness value in GA. Furthermore, generalized generation gap model and a modified SPX operator are utilized to increase the performance of real-coded genetic algorithm (RCGA).

The remaining of this paper is organized as follows: firstly, the formulation of groundwater management model (simulation model and optimization model) is described; secondly, a modified Pareto based Real-Coded Genetic Algorithm (mPRCGA) with generalized generation gap model and a modified SPX operator is proposed; thirdly, performance of the proposed mPRCGA based management model is tested on a hypothetical example.

2. Methodology

The main purpose of groundwater management model is to determine an optimal management strategy that maximizes the hydraulic, economic, or environmental benefits. Two sets of variables (decision variables and state variables) are involved, and the management strategies are usually constrained by some physical factors including well capacities, hydraulic heads, or water demand requirements. A groundwater management model is coupled with two main parts: simulation model and optimization model.

2.1. Formulation of Groundwater Simulation Model

The simulation model is the principal part of groundwater management model, since its solution is necessary in predicting the hydraulic response of aquifer system for different management strategies. The three-dimensional groundwater flow equation may be given as:

$$\frac{\partial}{\partial x_i}\left(K_{ij}\frac{\partial h}{\partial x_j}\right) + W = S_s \frac{\partial h}{\partial t} \quad i,j = 1,2,3 \tag{1}$$

where K_{ij} is the hydraulic conductivity tensor [L·T^{-1}], h is the hydraulic head [L], S_s is the specific storage [L^{-1}], t is time [T], W is the volumetric flux per unit volume (positive for inflow and negative for outflow) [T^{-1}], and x_i are the Cartesian coordinates [L].

In this study, the computer model of MODFLOW [27] is used to simulate the groundwater flow process.

2.2. Formulation of Groundwater Optimization Model

The optimization model is also absolutely necessarily for groundwater management models. In a groundwater optimization problem, the often-used objective is to maximize the total pumping or to minimize the total cost of capital, well drilling/installing and operating at a fixed demand. In this study, we use the minimization of total pumping cost as the objective of optimization model.

The objective function consists of capital cost, cost of well drilling/installing, and operating costs. Decision variables are pumping rates of candidate wells. The constraint set include some physical factors such as well capacities, hydraulic heads, or water demand requirements. The optimization model can be given as follows:

$$\min\left\{ a_1 \sum_{i=1}^{N} y_i + a_2 \sum_{i=1}^{N} y_i d_i + a_3 \sum_{i=1}^{N}\sum_{j=1}^{T} \left|Q_i^j\right|\left(H_i^0 - h_i^j\right)\right\} \tag{2}$$

subject to,

$$\sum_{i=1}^{N} \left|Q_i^j\right| \geq \widetilde{Q}^j, \quad j = 1,2,\cdots,T \tag{3a}$$

$$h_i^j \geq h_{i,\min}^j, \quad i = 1,2,\cdots,N; j = 1,2,\cdots,T \tag{3b}$$

$$Q_{i,\min}^j \leq Q_i^j \leq Q_{i,\max}^j, \quad i = 1,2,\cdots,N; j = 1,2,\cdots,T \tag{3c}$$

$$y_i = \begin{cases} 1 \text{ if } \sum_{j=1}^{T} Q_i^j \neq 0 \\ 0 \text{ if } \sum_{j=1}^{T} Q_i^j = 0 \end{cases}, \quad i = 1,2,\cdots,N; j = 1,2,\cdots,T \tag{3d}$$

where a_1 is the fixed capital cost per well in terms of dollars or other currency units [\$], a_2 is the installation and drilling cost per unit depth of well bore [\$/L], a_3 is the pumping costs per unit volume of flow [\$/L^3], y_i is a binary variable equal to either 1 if ith well is active or zero if ith well is inactive, d_i is the depth of well bore of ith well [L], $h_{i,\min}^j$ is the minimum hydraulic head value at ith well at time j [L], $Q_{i,\min}^j$ and $Q_{i,\max}^j$ are the ranges of allowable pumping rates for ith well at time j [L^3·T^{-1}], $\widetilde{Q}^j$ is the water demand at time j [L^3·T^{-1}], H_i^0 is the land surface elevation at ith well.

2.3. A Modified Pareto Dominance Based Genetic Algorithm (mPRCGA)

In this section, a modified Pareto dominance based real-coded genetic algorithm (mPRCGA) is proposed. The main features of mPRCGA are as: 1) vector combination of objective function and the total degree of constraint violation is preferred to weight combination; 2) Pareto intensity number is substituted for individual's fitness; 3) real-coded representation is used in GA; 4) generalized generation gap model (G3 model) is adopted as the population-alternation model; 5) modified SPX operator is used as recombination operator. The details of mPRCGA are described and explained below.

Step 1: Problem initialization and setting mPRCGA parameters

Let $f(x)$ be an objective function to be minimized, N be the number of decision variables, x_i be the ith decision variable to be determined $(i = 1, 2, \cdots, N)$, x be the vector $(x_1, x_2, \cdots, x_N)^{\mathrm{T}}$, and T is the transpose operator. Based on these definitions, the mathematical optimization problems can be stated as follows:

$$\min f(x), \quad \text{subject to} \quad x_i \in [l_i, u_i], i = 1, \cdots, N \tag{4}$$

where l_i and u_i are lower and upper bounds of the decision variables. In addition, there are M constraints including inequality constraints (g_1) and equality constraints (g_2) in the constrained optimization problem:

$$\begin{cases} g_{1j}(xx) \leq 0, \ j = 1, \cdots, q \\ g_{2j}(xx) = 0, \ j = q+1, \cdots, M \end{cases} \tag{5}$$

where q is the number of inequality constraints and M-q is the number of equality constraints.

To solve this optimization problem using mPRCGA, the constraints in Equation (5) should be converted into objective function. Vector combination of objective function and the total degree of constraint violation is used as follows:

$$\min(f(x), r(x)), \quad \text{subject to} \quad x_i \in [l_i, u_i], i = 1, \cdots, N \tag{6}$$

where $(f(x), r(x))$ is the vector composed of $f(x)$ and $r(x)$. $r(x)$ is the total degree of constraint violation, and can be obtained according to Equation (7) and Equation (8).

$$r(x) = \sum_{j=1}^{M} w_j v_j(x) \tag{7}$$

$$v_j(xx) = \begin{cases} \max\{0, g_{1j}(xx)\}, & 1 \leq j \leq q \\ |g_{2j}(xx)|, & q+1 \leq j \leq M \end{cases} \tag{8}$$

where w_j is weighing of jth constraint, $v_j(x)$ is the degree of jth constraint violation.

In this step the parameter sets of mPRCGA should also be defined: n_{pop} (population size), $Iter_{\max}$ (maximum generation), λ (parameters for G3 model), ε (expanding rate in SPX operator), c (parameter for Gaussian mutation in modified SPX operator).

Step 2: Generation of initial population

Make n_{pop} real-number vectors randomly and let them be an initial population P_t ($t = 0$).

Step 3: Individual ranking in population

As shown in Equation (6), the objective function is not a scalar but a vector. Thus, Pareto dominance is used to compare the vector [28]. On the basis of the vector comparison, Pareto intensity number is adopted to rank the individual in the population. The Pareto intensity number can be obtained as follows:

$$SI(i) = \#\left\{ x_j \big| \left(x_j \in P_t \right) \& \left(x_i \prec x_j \right) \right\} \tag{9}$$

where $SI(i)$ is the Pareto intensity number of ith individual in the population, P_t is the population in generation t, $x_i \prec x_j$ means x_i Pareto dominate x_j, # is cardinality of the set.

Step 4: Population improvement and updating

Population-alteration models and recombination operators are of great significance to real-coded GAs' performance. Generalized Generation Gap model (G3 model) is modified from MGG model and it is more computationally faster by replacing the roulette-wheel selection with a block selection of the best two solutions [29] [30].

UNDX and SPX are the most commonly used recombination operators. The UNDX operator uses multiple parents and Gaussian mutation to create offspring solutions around the center of mass of these parents. A small probability is assigned to solutions away from the center of mass. On the other hand, the SPX operator assigns a uniform probability distribution for creating offspring in a restricted search space around the region marked by the parents.

A modified SPX operator below is the combination of UNDX and SPX and can overcome some of their shortcomings. For simplicity, considering a 3-parent SPX in a two dimensional searching space as shown in **Figure 1**, where $x^{(1)}$, $x^{(2)}$ and $x^{(3)}$ are parent vectors, o is the center of the three parents. The inner triangle is formed by the three parent vectors firstly, then to be expanded to form the outer triangle. The vertex (of a triangle) is calculated as follows:

$$y^{(i)} = (1 + \varepsilon)\left(x^{(i)} - o \right) \tag{10}$$

where ε is the expanding rate. Thus a simplex is accomplished.

Then, the Gaussian mutation borrowed from UNDX operator is performed as follows:

$$x^{**} = x^* + r \sum_{i=1}^{N} \zeta_i e_i \tag{11}$$

where $e_i \left(i = 1, \cdots, N \right)$ is the unit coordinate vectors; r is the mean value of distances between each parent vector and center o; $\zeta_i \left(i = 1, \cdots, N \right)$ is zero-mean normally distributed variables with variance σ_ζ^2. Zhou *et al.* (2003) suggested $\sigma_\zeta = c/\sqrt{N}$ and observed that $c = 1$ to 1.3 performed well.

In Step 4, G3 model is employed as the main process, and the modified SPX is embedded and used as a sub process. Detailed process is as follows:

4a: Select $\mu(= n + 1)$ parents (best parent and μ-1 other parents randomly) from population P_t; Repeat (4b) λ times.

4b: Modified SPX procedure

4b.1: From the chosen μ parents to compute their center;

4b.2: Construct a simplex spanned by the chosen μ parents and its center;

4b.3: Select a point x^* randomly in the spanned simplex;

4b.4: Perform Gaussian mutation at point x^* to create an offspring x^{**}.

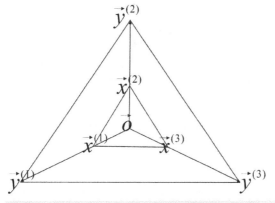

Figure 1. Illustration of a three-parent SPX operator.

4c: Choose two parents randomly from population P_t;

4d: Combine the randomly selected two parents (4c) and λ created offspring (4b) to form a population S;

4e: Rank individual of population S, choose the best two individuals;

4f: Replace the chosen two parents (4c) with these two individuals to update P_t.

Step 5: Repeat the above procedure from Step 3 to Step 4 until a certain stop criteria is satisfied.

3. Numerical Example

The performance of the mPRCGA based management model is tested on a hypothetical example considering multiple management periods.

3.1. Description

The example is to deal with the minimization of pumping cost from an unconfined aquifer system and it is assumed that the numbers and locations of the candidate wells are known. This example was previously solved using DDP (Differential Dynamic Programming) by Jones *et al.* (1987), GA and SA by Wang and Zheng (1998), and HS (Harmony Search algorithm) by Ayvaz (2009). **Figure 2** shows the plan view of the aquifer system under consideration.

Groundwater is pumped from an unconfined aquifer with a hydraulic conductivity of 86.4 m/day and specific yield of 0.1. As can be seen from **Figure 2**, boundary conditions of the aquifer include the Dirichlet boundary at the north and no-flow at the other sides. The distance between land surface and aquifer bottom is 150 m. The flow model is transient; it is assumed that initial hydraulic head value is 100 m everywhere.

3.2. Optimization Model

The total management period is one year, which is divided into four stress periods of 91.25 days each. There are eight candidate pumping wells, and the water demands for each period are 130,000, 145,000, 150,000, and 130,000 m³/day, respectively. The hydraulic head must above zero (bottom) anywhere in the aquifer, and each pumping rates must be in the range of 0 to 30,000 m³/day. The objective function to be minimized is in the form of Equation (2) with $T = 4$. Note the first two terms in Equation (2) is neglected and Equation (2) is reduced to the last term.

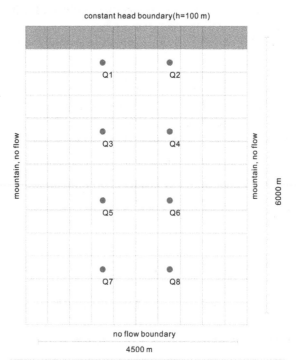

Figure 2. Plan view of unconfined aquifer model.

3.3. Results and Discussion

Using the parameter sets given in **Table 1**, the optimum pumping rates and total cost has been solved through the proposed mPRCGA based management model. **Table 2** summarizes the results of mPRCGA as well as other studies.

Table 1. Parameter sets in mPRCGA.

n_{pop}	$Iter_{max}$	λ	ε	c
500	1000	50	5	1.15

Table 2. Comparison of optimum pumping rates and total cost for example 1 by different optimization methods (Unit: m³/d and $).

Well	Period	DDP	GA	SA	HS	mPRCGA
1	1	30,000	28,000	30,000	29,476	30,000
2		30,000	28,000	30,000	29,458	30,000
3		21,924	28,000	17,000	18,600	21,503
4		21,924	4000	17,000	20,597	21,503
5		7494	12,000	16,000	10,764	7703
6		7494	12,000	11,000	12,801	7703
7		5582	14,000	7000	2452	5794
8		5582	,000	2000	5853	5794
demand		130,000	130,000	130,000	130,000	130,000
1	2		28,000	30,000	29,919	30,000
2			28,000	30,000	29,721	30,000
3			10,000	21,000	25,125	27,523
4			20,000	21,000	19,792	27,523
5			8000	10,000	11,482	8681
6			14,000	12,000	13,757	8681
7			16,000	12,000	7865	6296
8			20,000	9000	7339	6296
demand			144,000	145,000	145,000	145,000
1	3		28,000	30,000	29,795	30,000
2			30,000	30,000	29,599	30,000
3			12,000	25,000	24,188	27,236
4			28,000	18,000	27,917	27,236
5			6000	12,000	12,160	9803
6			8000	15,000	12,947	9803
7			26,000	10,000	3299	7961
8			12,000	10,000	10,096	7961
demand			150,000	150,000	150,000	150,000
1	4		28,000	30,000	29,876	30,000
2			28,000	30,000	29,976	30,000
3			20,000	22,000	20,200	21,357
4			22,000	12,000	22,185	21,357
5			4000	5000	10,902	7669
6			8000	13,000	10,506	7669
7			6000	6000	160	5974
8			14,000	12,000	6196	5974
demand			130,000	130,000	130,000	130,000
Total cost ($)		28,693,336	29,779,432	29,572,110	29,540,860	29,539,705
Number of simulation		4	~375,000	~450,000	256,615	50,500

This shows that the results of mPRCGA are in good agreement with the water demands for each period. Also, pumping rates of wells near the Dirichlet boundary condition are higher than other wells (Q1 = Q2 > Q3 = Q4 > Q5 = Q6 > Q7 = Q8), as expected. The results obtained by mPRCGA also satisfies the requirement of symmetry of aquifer system, this also verify the reliability of MPRCGA.

Furthermore, by comparison of total pumping cost, MPRCGA finds better objective function value (29,539,705 $) than GA (29,779,432 $), SA (29,572,110 $), and HS (29,540,860 $). DDP gives a better result (28,693,336 $), may be due to the stage-wise decomposition of the algorithm. However, the realization of DDP is much more complex. The contrast in pumping rates in 1st period shows that mPRCGA gives closer results to DDP than other methods, and the maximum relative deviation of single well pumping rates is less than 4%. This indicates that proposed mPRCGA is a reliable and effective heuristic optimization method.

Finally, the number of simulations (in **Table 2**) shows that mPRCGA requires 50,500 simulations, whereas GA, SA, and HS require ~375,000, ~450,000, and 256,615 simulations, respectively.

4. Summary and Conclusions

This study proposes a groundwater management model in which the solution is performed through a combined simulation-optimization model. MODFLOW is used as the simulation model, and in the optimization model mPRCGA approach is proposed. In this approach, a constraint handling technique is presented based on the multi-objective concepts and Pareto dominance is introduced to compute the Pareto intensity number of individual to substitute for the conventional fitness value; meanwhile, RCGA is modified by adopting generalized generation gap model and a modified SPX operator. A hypothetical example is utilized to test the accuracy of mPRCGA. The results indicate that the proposed mPRCGA is an effective method for solving the groundwater management models. Some major conclusions can be drawn as follows:

- The proposed mPRCGA can be applied for groundwater management problems, compared with other optimization approaches, mPRCGA can obtain satisfactory results as shown in **Table 2**;
- The parameter sets in mPRCGA, shown in **Table 1**, is more easily obtained and realized than penalty function methods, this advantage tends to greatly enhance the robustness of algorithm;
- To some extent, the modified SPX operator partially overcome the limitation of offspring generation of SPX operator and UNDX operator, together with generalized generation gap model, the modified RCGA may have good ergodic property and high probability to find the global optimum solution.

Acknowledgements

This study was supported by Science and Technology Commission of Shanghai Municipality (No.12231200700).

References

[1]   Ayvaz, M.T. (2009) Application of Harmony Search Algorithm to the Solution of Groundwater Management Models. *Advances in Water Resources*, **32**, 916-924. http://dx.doi.org/10.1016/j.advwatres.2009.03.003

[2]   Ayvaz, M.T. and Karahan, H. (2008) A Simulation/Optimization Model for the Identification of Unknown Groundwater Well Locations and Pumping Rates. *Journal of Hydrology*, **357**, 76-92. http://dx.doi.org/10.1016/j.jhydrol.2008.05.003

[3]   Doughterty, D.E. and Marryott, R.A. (1991) Optimal Groundwater Management. 1 Simulated Annealing. *Water Resources Research*, **27**, 2493-2508. http://dx.doi.org/10.1016/j.jhydrol.2008.05.003

[4]   Ahlfeld, D.P., Barlow, P.M. and Mulligan, A.E. (2005) GWM-A Ground-Water Management Process for the US Geological Survey Modular Ground-Water Model (MODFLOW-2000). US Geological Survey Open-File Report

[5]   Mckinney, D.C. and Lin, M.D. (1994) Genetic Algorithm Solution of Groundwater Management Models. *Water Resources Research*, **30**, 1897-1906. http://dx.doi.org/10.1029/94WR00554

[6]   Cheng, A.H.D., Halhal, D., Naji, A. and Ouazar, D. (2000) Pumping Optimization in Saltwater-Intruded Coastal Aquifers. *Water Resources Research*, **36**, 2155-1265. http://dx.doi.org/10.1029/2000WR900149

[7]   Culver, T.B. and Shoemaker, C.A. (1992) Dynamic Optimal Control for Groundwater Remediation with Flexible Management Periods. *Water Resources Research*, **28**, 629-641. http://dx.doi.org/10.1029/91WR02826

[8]   Hallaji, K. and Yazicigil, H. (1996) Optimal Management of Coastal Aquifer in Southern Turkey. *Journal of Water*

Resources Planning and Management, **122**, 233-244. http://dx.doi.org/10.1061/(ASCE)0733-9496(1996)122:4(233)

[9]   Jones, L., Willis, R. and Yeh, W.W.G. (1987) Optimal Control Nonlinear Groundwater Hydraulics Using Differential Dynamic Programming. *Water Resources Research*, **23**, 2097-2106. http://dx.doi.org/10.1029/WR023i011p02097

[10] Wang, M. and Zheng, C. (1998) Ground Water Management Optimization Using Genetic Algorithms and Simulated Annealing: Formulation and Comparison. *Journal of the American Water Resources Association*, **34**, 519-530. http://dx.doi.org/10.1111/j.1752-1688.1998.tb00951.x

[11] Sun, N. (1999) Inverse Problems in Groundwater Modeling. Kluwer Academic Publishers, Dordrecht. http://dx.doi.org/10.1007/978-94-017-1970-4

[12] Cieniawski, S.E., Eheart, J.W. and Ranjithan, S. (1995) Using Genetic Algorithm to Solve a Multiobjective Groundwater Monitoring Problem. *Water Resources Research*, **31**, 399-409. http://dx.doi.org/10.1029/94WR02039

[13] Wu, J., Zhu, X. and Liu, J. (1999) Using Genetic Algorithm Based Simulated Annealing Penalty Function to Solve Groundwater Management Model. *Science in China Series E-Technological Sciences*, **42**, 521-529.

[14] Mahinthakumar, G.K. and Sayeed, M. (2005) Hybrid Genetic Algorithm—Local Search Methods for Solving Groundwater Source Identification Inverse Problems. *Journal of Water Resources Planning and Management*, **131**, 45-57. http://dx.doi.org/10.1061/(ASCE)0733-9496(2009)135:5(314)

[15] Bhattacharjy, R.K. and Datta, B. (2009) ANN-GA-Based Model for Multiple Objective Management of Coastal Aquifers. *Journal of Water Resources Planning and Management*, **135**, 314-322. http://dx.doi.org/10.1061/(ASCE)0733-9496(2009)135:5(314)

[16] Wang, Y., Cai, Z., Zhou, Y. and Xiao, C. (2009) Constrained Optimization Evolutionary Algorithms. *Journal of Software*, **20**, 11-29 (In Chinese). http://dx.doi.org/10.3724/SP.J.1001.2009.00011

[17] Michalewicz, Z. and Schoenauer, M. (1996) Evolutionary Algorithms for Constraint Parameter Optimization Problems. *Evolutionary Computation*, **4**, 1-32. http://dx.doi.org/10.1162/evco.1996.4.1.1

[18] Guan, J., Kentel, E. and Aral, M.M. (2008) Genetic Algorithm for Constrained Optimization Models and Its Application in Groundwater Resources Management. *Journal of Water Resources Planning and Management*, **134**, 64-72. http://dx.doi.org/10.1061/(ASCE)0733-9496(2008)134:1(64)

[19] Keshari, A.K. and Datta, B. (2001) A Combined Use of Direct Search Algorithms and Exterior Penalty Function Method for Groundwater Pollution Management. *Journal of Porous Media*, **4**, 259-270.

[20] Ricciardi, K.L., Pinder, G.F. and Karatzas, G.P. (2007) Efficient Groundwater Remediation System Design Subject to Uncertainty Using Robust Optimization. *Journal of Water Resources Planning and Management*, **133**, 253-263. http://dx.doi.org/10.1061/(ASCE)0733-9496(2007)133:3(253)

[21] Cai, Z. and Wang, Y. (2006) A Multiobjective Optimization-Based Evolutionary Algorithm for Constrained Optimization. *IEEE Transactions on Evolutionary Computation*, **10**, 658-675. http://dx.doi.org/10.1109/TEVC.2006.872344

[22] Deb, K. (2000) An Efficient Constraint Handling Method for Genetic Algorithms. *Computer Methods in Applied Mechanics and Engineering*, **186**, 311-338. http://dx.doi.org/10.1016/S0045-7825(99)00389-8

[23] Coello, C.A.C. (2000) Constraint Handling Using an Evolutionary Multiobjective Optimization Technique. *Civil Engineering and Environmental Systems*, **17**, 319-346. http://dx.doi.org/10.1080/02630250008970288

[24] Venkatraman, S. and Yen, G.G. (2005) A Generic Framework for Constrained Optimization Using Genetic Algorithm. *IEEE Transactions on Evolutionary Computation*, **9**, 424-435. http://dx.doi.org/10.1109/TEVC.2005.846817

[25] Zhou, Y., Li, Y., Wang, Y. and Kang, L. (2003) A Pareto Strength Evolutionary Algorithm for Constrained Optimization. *Journal of Software*, **14**, 1243-1249. (In Chinese)

[26] Mezura-Montes, E.E. and Coello, C.A.C. (2005) A Simple Multimembered Evolution Strategy to Solve Constrainted Optimization Problems. *Evolutionary Computation*, **9**, 1-17.

[27] McDonald, M.C. and Harbough, A.W. (1988) A Modular Three-Dimensional Finite Difference Groundwater Flow Model. US Geological Survey Book 6, Chapter A1.

[28] Van Veldhuizen, D.A. and Lamont, G.B. (2000) Multiobjective Evolutionary Algorithms: Analyzing the State-of-the-Art. *Evolutionary Computation*, **8**, 125-147. http://dx.doi.org/10.1162/106365600568158

[29] Deb, K., Anand, A. and Joshi, D. (2002) A Computationally Efficient Evolutionary Algorithm for Real-Parameter Optimization. *Evolutionary Computation*, **10**, 371-395. http://dx.doi.org/10.1162/106365602760972767

[30] Thangavelu, S. and Velayutham, C. (2010) Taguchi Method Based Parametric Study of Generalized Generation Gap Genetic Algorithm Model. In: Panigrahi, B., Das, S., Suganthan, P. and Dash, S., Eds., *Swarm, Evolutionary, and Memetic Computing*, Springer, Berlin/Heidelberg, 344-350. http://dx.doi.org/10.1007/978-3-642-17563-3_42

6

Influence of Potential Evapotranspiration on the Water Balance of Sugarcane Fields in Maui, Hawaii

Javier Osorio[1], Jaehak Jeong[1], Katrin Bieger[1], Jeff Arnold[2]

[1]Blackland Research and Extension Center, Texas A & M AgriLife Research, Temple, USA
[2]Grassland, Soil and Water Research Laboratory, United States Department of Agriculture, Agricultural Research Service (USDA-ARS), Temple, USA
Email: josorio@brc.tamus.edu

Abstract

The year-long warm temperatures and other climatic characteristics of the Pacific Ocean Islands have made Hawaii an optimum place for growing sugarcane; however, irrigation is essential to satisfy the large water demand of sugarcane. Under the Hawaiian tropical weather, actual evapotranspiration (AET) is the primary mechanism by which water is removed from natural and agricultural systems. The Hawaiian Commercial and Sugar Company (HC&S), the largest sugarcane grower of the Hawaiian Islands, has developed a locally optimized AET equation for the purpose of water management on its 184.3 km² sugarcane plantation on the Island of Maui. In this paper, in order to assess the influence of AET on the hydrological water balance of the HC&S' sugarcane cropping system, the performance of the HC&S method was compared with three physically-based methods: Penman-Monteith, Priestley-Taylor, and Hargreaves, as well as, to a set of historical pan evaporation data. A Soil and Water Assessment Tool (SWAT) project was setup to estimate the water balance in two sugarcane fields: a windy lowland field and a rocky highland field on a hill slope. Under Hawaiian weather conditions, wind speed was found to be the most influential climatic parameter over potential evapotranspiration (PET); therefore, the results with both Hargreaves and Priestley-Taylor underpredicted PET by approximately 30%, presumably because these methods do not take wind speed into account. The HC&S method was demonstrated to be the most accurate PET method compared to the other commonly used PET equations, with less than 10% error. Of the annual total water supply of 3400 mm, AET accounted for 75% - 80% of the total water consumption. These findings can be used to improve the irrigation efficiency as well as other management scenarios to optimize water use on the Island of Maui.

Keywords

Evapotranspiration, Water Balance, Hydrological Modeling, Sugarcane, SWAT

1. Introduction

The year-long warm temperatures and other climatic characteristics of the Pacific Ocean Islands have made the Island of Maui an optimum place for growing sugarcane (*Saccharum officinarum* L.). Because of the high water demand of sugarcane, irrigation becomes one of Maui's major constraints [1]. According to McMahon *et al.* [2] irrigation is essential to satisfy the large water demand of sugarcane for optimum growth and yield. In Maui, large quantities of surface and ground water from the windward side of the island are diverted by intake structures into ditches and tunnels to transport the water to the leeward plains, where the best agricultural soils are located [1]. However, mild droughts and periods of low rainfall have adversely affected the perennial streams and depleted high-level groundwater aquifers that supply Maui's irrigation system. As water resources become limited, effective use of available water for irrigation is of concern throughout the sugarcane industry [3].

The Hawaiian Commercial & Sugar Company (HC&S), the largest sugarcane grower of the Hawaiian Islands, has an area of 184.3 km^2 under sugarcane production on the Island of Maui. The productivity of sugarcane on Maui (159.9 t·ha^{-1} sugarcane yield and 24.9 t·ha^{-1} sugar yield) is known to be above the national average (78.8 t·ha^{-1} sugarcane yield and 9.2 t·ha^{-1} sugar yield) and higher than other major sugarcane producing regions in USA like Louisiana (51.3 t·ha^{-1} sugarcane yield and 6.6 t·ha^{-1} sugar yield) or Texas (85.9 t·ha^{-1} sugarcane yield and 10.0 t·ha^{-1} sugar yield) [4]. The sugarcane cropping period is approximately 24 months. The company requires 760,000 m^3·d^{-1} of fresh water for irrigation, which is obtained from surface (fog drip and rainfall) and ground water sources (16 deep brackish-water wells) and is applied directly to the root zone of the sugarcane plants through a drip irrigation system.

From the hydrological perspective, actual evapotranspiration (AET) is the primary mechanism by which water is removed from the hydrologic cycle [5]-[7]. AET returns land-based water to the atmosphere through processes such as transpiration, evaporation from the plant canopy and the soil and sublimation when snow is present [8]. The rate of AET for a given environment is a function of four critical factors: soil moisture, plant type, stage of plant development, and weather conditions. In addition, the transpiration rate is influenced by soil management practices [7] [9]. McCuen [10] asserts that the most influential climatic factors on ET are temperature, relative humidity, radiation rates and wind speed.

Large AET rates in agricultural regions impact water quantity and quality [11] [12]. Accordingly, the amount of water needed for irrigation of agricultural systems is frequently calculated based on estimated AET. Underestimation of AET leads to an insufficient water supply to the crop and consequently to water stress, while overestimation of AET leads to a waste of water for irrigation. Therefore, accurate estimation of AET plays a critical role in the assessment, planning and management of water resources in general [11]. Several methods are used to estimate AET based on mass transfer, energy budget, water budget, soil moisture budget, ground-water fluctuations or meteorological variables. Each method has advantages and disadvantages; moreover, none is applicable under all conditions because of assumptions in the method or type of required data [13]. Most of the hydrological models employ the concept of potential evapotranspiration (PET) as the driving function for the calculation of AET [7] [14] [15]. PET is defined as the amount of water that would be removed from a vegetated landscape with no restrictions other than the atmospheric demand [16], and is often measured with evaporation pans. Also, PET can be estimated using methods based on available climate data [7] [17]. Currently, a number of methods that vary in complexity and data requirements are employed for calculating PET [2] [7] [9]. Their ability to produce consistent and meaningful PET estimates depends on their assumptions, data requirements and consideration of atmospheric factors [2] [15]. HC&S has remained in the sugarcane industry for more than 100 years and during this time the company has made numerous innovations on sugarcane technology, irrigation techniques, and crop management for the sugarcane plantation on the Island of Maui. To optimize water usage for irrigation purposes, HC&S has developed a locally calibrated PET method, which is based on pan evaporation measurements and is specific to sugarcane.

The main objective of this study was to evaluate the efficiency of the HC&S irrigation system by comparing the accuracy of AET estimation by the HC&S method against three physically-based methods: Penman-Monteith, Priestley-Taylor, and Hargreaves, as well as, to a set of historical pan evaporation data and to identify an AET method that is applicable to other crops and transferable to locations with similar characteristics.

2. Data and Methods

2.1. SWAT Model

SWAT is a physically based, semi-distributed, and long-term continuous watershed scale model that runs on a daily time step to predict the impact of climate, landuse, soil type, topographic characteristics, and land management practices on hydrology, sediment, nutrients, pesticides, and bacteria in large ungauged watersheds [18] [19]. SWAT estimates PET using physically based methods that are based on historical weather variables and plant/soil available water. The methods included are Hargreaves, Priestley-Taylor, and Penman-Monteith.

Runoff is estimated by the SCS Curve Number (CN) method (SCS, 1986) or the Green & Ampt method [20]. If the Green and Ampt method is used to calculate surface runoff, the model calculates canopy storage and evaporates any readily available water present in the canopy. In contrast, the CN method lumps canopy interception, surface storage and infiltration in its initial abstraction term. After calculating runoff, SWAT calculates the amount of transpiration under ideal conditions as a function of available water in the soil. Sublimation will occur if snow is present; otherwise, the actual amount of water evaporated from the soil is calculated as a function of soil depth, water content and the above-ground biomass.

SWAT simulates the water balance according to Equation (1), which includes AET, canopy interception, plant transpiration and soil evaporation, surface runoff, and vertical water movement in the unsaturated soil zone to the ground water:

$$SW_i = SW_0 + (R_i + I_i) - (Q_i + AET_i + PER_i) \tag{1}$$

where, SW_i is the final soil water content (mm) in day i, SW_0 is the initial soil water content (mm), i is the time (days), R_i is the amount of precipitation (mm·day^{-1}), I_i is the amount of daily irrigation (mm·day^{-1}), Q_i is the amount of surface runoff (mm·day^{-1}), AET_i is the amount of actual evapotranspiration (mm·day^{-1}), and PER_i is the amount of water percolating through the soil profile (mm·day^{-1}).

2.2. PET Methods

The Penman-Monteith method [21] (Equation (2)) is a theoretically based approach that incorporates energy and aerodynamic considerations [22]. It includes a measure of the resistance to the diffusion of water vapor into the Penman's equation, which is a radiation-aerodynamic combination equation to predict evaporation from open water, bare soil, and grass [23]:

$$\lambda E = \frac{\Delta \times (H_{net} - G) + \rho_{air} \times c_p \times (e_z^0 - e_z)/r_a}{\Delta + \gamma \times \left(1 + \frac{r_c}{r_a}\right)} \tag{2}$$

where: λ is the latent heat flux density (MJ·m^{-2}kg^{-1}), E is the depth rate evaporation (mm·day^{-1}), Δ is the slope of the saturation vapor pressure-temperature curve (kPa°C^{-1}), H_{net} is the net radiation (MJ·m^{-2}d^{-1}), G is the heat flux density to the ground (MJ·m^{-2}d^{-1}), ρ_{air} is the air density (kg·m^{-3}), c_p is the specific heat constant pressure (kPa·kg^{-1}°C^{-1}), e_z^0 is the saturation vapor pressure of air at height z (kPa), e_z is the water vapor pressure of air at height z (kPa), γ is the psychometric constant (kPa°C^{-1}), r_c is the plant canopy resistance (s·m^{-1}), and r_a is the diffusion resistance of the air layer (aerodynamic resistance) (s·m^{-1}).

The Priestley-Taylor method [24] (Equation (3)) is an energy-based approach. According to Jensen et al. [16] under wet conditions, evaporation from surfaces could be estimated using a simplified version of the Penman equation, in which the aerodynamic was deleted and the energy term is assumed to be a constant fraction $\alpha_0 = 1.28$.

$$\lambda E_0 = \alpha_0 \times \frac{\Delta}{\Delta + \gamma} \times (H_{net} - G) \tag{3}$$

where: λ is the latent heat of vaporization (MJ·kg^{-1}), E_0 is the PET (mm·d^{-1}), α_0 is a coefficient, Δ is the slope of the saturation vapor pressure-temperature curve (kPa°C^{-1}), γ is the psychometric constant (kPa°C^{-1}), H_{net} is the net radiation (MJ·m^{-2}d^{-1}), and G is the heat flux density to the ground (MJ·m^{-2}d^{-1}). In semiarid conditions this equation will underestimate PET.

The Hargreaves method [25] (Equation (4)) was originally derived from eight years of cool season Alta fescue grass lysimeter data from Davis, CA. According to Hargreaves and Allen [23] the main advantage of the Hargreaves approach to the other methods is the reduced data requirement.

$$\lambda E_0 = 0.002 \times H_0 \times \left(T_{max} - T_{min}\right)^{0.5} \times \left(\overline{T}_{av} + 17.8\right) \tag{4}$$

where: λ is the latent heat of vaporization (MJ·kg^{-1}), E_0 is the PET (mm·d^{-1}), H_0 is the extraterrestrial radiation (MJ·m^{-2}d^{-1}), T_{max} is maximum air temperature for a given day (°C), T_{min} is minimum air temperature for a given day (°C), and T_{av} is the mean air temperature for a given day (°C).

The HC&S method (Equation (5)) is specific for sugarcane and also site-specific to the central isthmus of Maui. It was developed to take into account that evapotranspiration is greatly modified by the marine surroundings and the topographic characteristics of the Island of Maui. The HC&S method to calculate PET is an empirical relationship that relates extensive pan evaporation data collected in the central isthmus of Maui [26] and locally measured meteorological parameters, such as air temperature, humidity, solar radiation and wind.

This method calculates PET by including a coefficient that multiplies the saturation vapor pressure, average actual vapor pressure, the slope of the saturation vapor pressure, a weighting function for temperature, net radiation expressed as equivalent evaporation and wind function.

$$E_0 = C \times W \times H_{net} + C \times \left(1 - W\right) \times \left(e_z^0 - e_z\right) \times V_f \tag{5}$$

where E_0 is the PET (mm·d^{-1}), C is a correction coefficient, H_{net} is the net radiation expressed in equivalent evaporation (mm·d^{-1}), W is a weighting function for temperature, e_z^0 is the saturation vapor pressure of air at average temperature (kPa), e_z is the actual water vapor pressure of the air (kPa), and V_f is a wind function.

The correction coefficient (C) is calculated with a polynomial equation that includes eight coefficients and the ratio of average diurnal and nocturnal wind speed, solar radiation expressed in equivalent evaporation, and the mean wind speed. PET is multiplied by a crop coefficient (K_c) to calculate AET. To account for seasonal differences in crop growth and AET, two different crop coefficient curves are used depending on the time of planting.

3. Study Area

The study area is located on the Island of Maui, Hawaii (20°54' latitude and 156°26' longitude) (**Figure 1**) and consists of sugarcane fields (184.3 km^2) that lie on elevations between sea level and 332 m. The climate of Maui is controlled primarily by topography and the position of the North Pacific anticyclone and other migratory weather systems relative to the island [11]. The climate is characterized by mild temperatures, cool and persistent trade winds, a rainy winter season from October through April, and a dry summer season from May through September [8] [27]. Solar energy and length of day are relatively uniform throughout the year and the surrounding ocean provides moist air and keeps temperatures fairly constant without extremes throughout the year [27]. Mean annual precipitation on Maui Island ranges from 7000 mm to 400 mm or less [28]. According to NOAA [29], based on a 30-year dataset from the Kahului Station, Maui's Central Valley, where the HC&S Co. is located, receives an average rainfall of 453 mm per year and has an average daily temperature of 24.3°C (ranges from 19.7°C to 29.0°C) and an average wind velocity of 5.7 m·s^{-1}. Relative humidity ranges from 53% to 81%.

Two fields were selected for this analysis: 905 and 415. They are located in contrasting micro-climates and landscapes. Field 905 is located at a lower elevation (31.0 m a.s.l.) with a slope of 4.3%. The soil texture is classified as silt-loam and the soils are in hydrologic group B. In contrast, Field 415 is located at a higher elevation (195.4 m a.s.l.) with a slope of 6.0%. The soil texture is classified as silty-clay-loam and the soil belongs to hydrologic group C [30]. Solar radiation (**Figure 2(a)**), wind velocity (**Figure 2(b)**), average temperature (**Figure 2(c)**) and relative humidity (**Figure 2(d)**) are plotted for both fields. There are no significant differences between fields 415 and 905 with regard to average temperature (23.7°C and 23.6°C), relative humidity (0.71 and 0.68), and solar radiation (18.9 and 20.0 MJ·m^{-2}). However, wind velocity shows significant differences between both fields (2.5 and 5.8 m·s^{-1}), with higher speeds found in Field 905. Relative differences for average temperature, relative humidity and solar radiation are: 2.8%, 5.9% and 2.2% respectively, while the relative difference for wind velocity is 54.9%. Temperatures and humidity are higher at Field 415 while solar radiation and wind speed

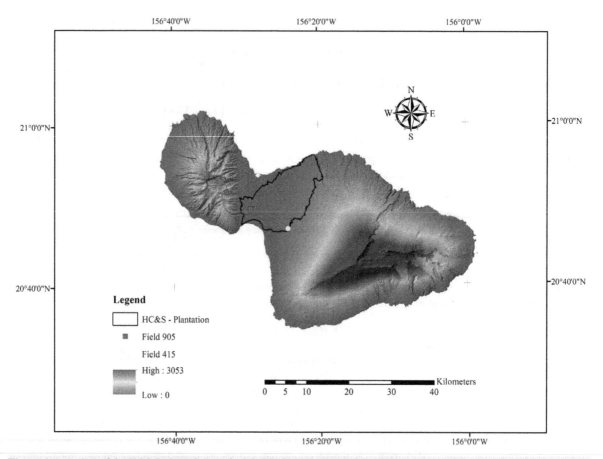

Figure 1. Location of the HC&S plantation on the Island of Maui, Hawaii.

are higher at Field 905. On Field 415, the sugarcane was planted in winter, while on Field 905 the crop was established in summer.

3.1. Input Data and Model Setup

The Digital Elevation Model (DEM) with a resolution of 10 m (1/3 arc-second) was acquired from the National Elevation Dataset (NED) [31] (available at http://viewer.nationalmap.gov/viewer/). Detailed soil map and properties for the study area were obtained from the Natural Resources Conservation Service (NRCS)—Soil Survey Geographic (SSURGO) Database (Soil Survey Staff NRCS-USDA, 2011) available at http://soildatamart.nrcs.usda.gov. The soils in the HC&S plantation area are classified as Mollisols (46.3%), Andisols (15.4%) and Aridisols (15.1%), Oxisols (10.4%), Entisols (6.0%), Inceptisols (4.2%) and Ultisols (2.3%). The landuse map and related information were obtained from the United States Geological Service (USGS) Land Cover Institute (LCI) [32] available at the USGS-Seamless Data Warehouse (http://seamless.usgs.gov/) with a resolution of 30 m (1 arc-second). The agricultural land was characterized as a 2-year sugarcane crop. The ArcGIS Interface for SWAT2012 was used to delineate the boundaries of 701 sugarcane plots, defining each plot as an individual Hydrologic Response Unit (HRUs) with sizes ranging from 1.4 to 166.0 ha.

The model uses a 12-year dataset of daily climatic records from 39 rain gauges and 12 weather stations maintained by HC&S. The climatic variables included are precipitation, minimum and maximum temperature, relative humidity, solar radiation and wind speed. Missing values in the dataset were simulated by the model's built-in weather generator based on the weather statistics of the study area. HC&S applies water to the sugarcane fields through a drip irrigation system. The company maintains records of the applied volumes of water and the date of irrigation for each individual field. The SWAT model source code was modified to automatically read a database of dates and measured volumes of water applied to each field for irrigation purposes. For this study, historical daily irrigation data from 2003 to 2012 were acquired from the HC&S archive. Based on data availa-

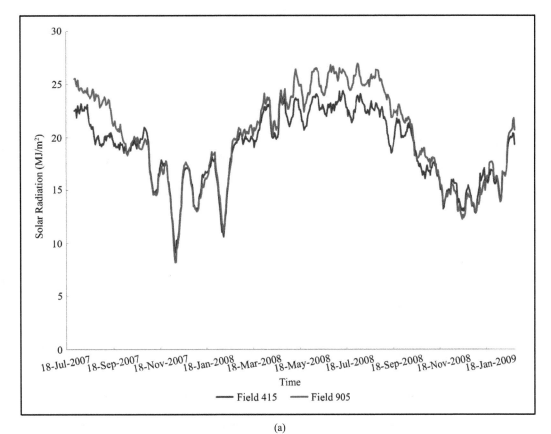

(a)

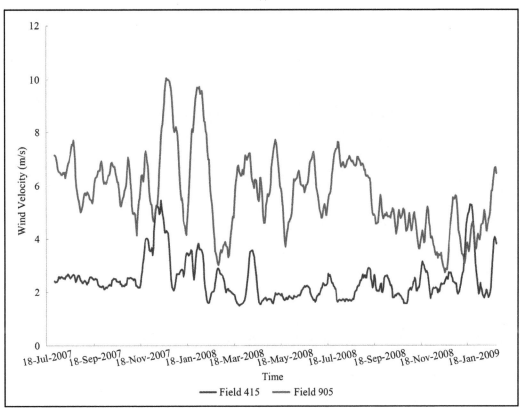

(b)

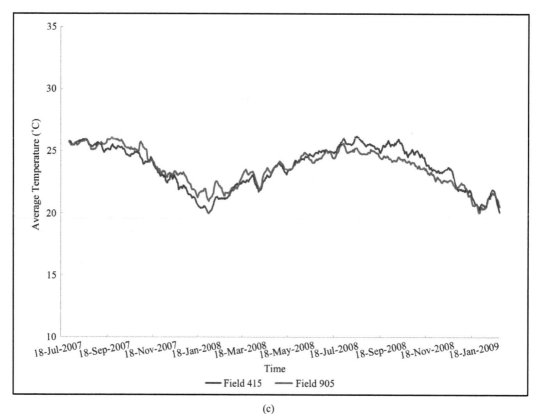

(c)

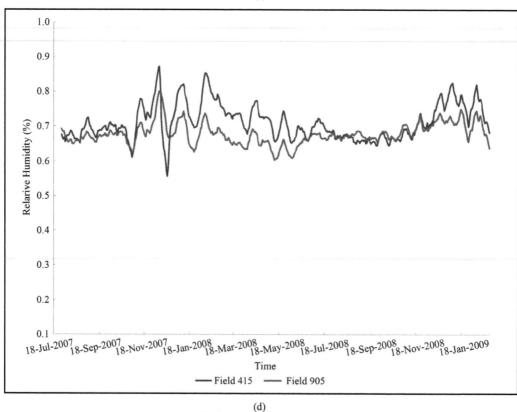

(d)

Figure 2. Measured climatic variables for the crop period at fields 415 and 905: (a) Solar radiation (MJ·m^{-2}); (b) Wind velocity (m·s^{-1}); (c) Average temperature (°C); and (d) Relative humidity (%).

bility, the simulation period was set up for 10 years (2003-2012) with the first year of simulation set as a spin-up period. However, the analysis of model output was performed for the crop period that comprises the years 2007 through 2009.

The water balance analysis was done evaluating surface runoff, percolation, lateral flow, precipitation, irrigation and available soil water. In addition, the relative difference between the predicted average annual PET and measured pan evaporation was calculated for the four PET methods included in this study. Measurements of pan evaporation in Hawaii began in 1894 and include intermittent observations for over 200 sites [26]. Most sites had standard, galvanized Class A pans on the surface, but many were later replaced with stainless steel pans to avoid corrosion and set on 1.52 m (equivalent to 5 ft) high platforms. The pan evaporation data used in this study ware corrected to account for differences due to height and pan material. A total of 12 galvanized Class A pans provided a data set with accumulated monthly and annual evaporation for the interval between 1963 and 1983. There were no Pan A measurements at the fields evaluated in this study; therefore, the closest available information to each field was used for comparison.

3.2. Crop Yield Calibration

The water balance in agricultural watersheds is influenced by crop production [33]. Crop dry matter production in Hawaii depends on the amount of water available from rainfall and irrigation [34]. The yield of a given crop can generally be described as a function of cumulative AET [35]. The AET has been shown to be directly related to dry biomass production when factors such as fertility, sunshine, temperature, and soil moisture are not limiting [36]. The following SWAT parameters were used for the purpose of manually calibrating crop yield: the soil evaporation compensation coefficient (ESCO), the plant uptake compensation factor (EPCO) and the biomass-energy ratio (BIO_E) ($kg \cdot ha^{-1}/MJ \cdot m^{-2}$). ESCO adjusts the depth distribution of soil evaporation to meet soil evaporative demand and varies between 0.01 and 1.0. As the value of ESCO is reduced, the model is able to evaporate more water from deeper layers in the soil profile. EPCO compensates water between soil horizons to meet the potential water uptake and ranges from 0.01 to 1.0. As EPCO approaches 1.0, the model allows more of the water uptake demand to be met by lower layers in the soil. BIO_E is the amount of dry biomass produced per unit intercepted solar radiation in ambient CO_2 and varies between 10 and 90. The greater BIO_E, the greater the potential increase in total plant biomass on a given day. An important parameter governing crop growth in SWAT is the leaf area index (LAI), which controls the amount of light intercepted by the plant and converted to biomass. Also, LAI is used to calculate AET. When LAI is larger than 3, SWAT assumes that AET equals PET. To calibrate the LAI curve calculated by SWAT, it was compared to the crop coefficient curve used by the HC&S method. To evaluate crop yields simulated by SWAT, they were compared to field measurements of crop yield obtained between 2003 and 2012.

4. Results

4.1. Calibration of Sugarcane Yield

While the default values for the parameters ESCO and EPCO of 0.95 and 1.0 and BIO_E ($kg \cdot ha^{-1}/MJ \cdot m^{-2}$) of 25 were used to initialize the model; the calibrated values for the same parameters were 0.5, 0.98, and 22.5, respectively. By calibrating BIO_E, the leaf area index curve break point (LAI = 3) was adjusted to the break point of the crop coefficient ($K_c = 1$) used by HC&S for both winter (Field 415) and summer (Field 905) planting (**Figure 3(a)** and **Figure 3(b)**). Simulated average crop yields for the 2003-2012 period of time are 71.7 $t \cdot ha^{-1}$ (± 11.3 $t \cdot ha^{-1}$) and 77.7 $t \cdot ha^{-1}$ (± 12.0), for fields 415 and 905 respectively. Field measurements of crop yield for fields 415 and 905 provide an average value of 71.3 $t \cdot ha^{-1}$ (± 7.7 $t \cdot ha^{-1}$) and 77.3 $t \cdot ha^{-1}$ (± 10.0 $t \cdot ha^{-1}$), respectively. Predicted crop yields were not different from measured values for both fields (relative differences lower than 1%). No other goodness-of-fit metrics were used in this study due to limitations in measured data. Successful calibration of crop yield guaranteed that AET was adequately represented by the model and varied accordingly as a function of climatic variables and crop development. In addition, properly predicted AET assured appropriate partitioning of the remaining components of the water budget (**Figure 3(a)**, **Figure 3(b)**).

4.2. Potential Evapotranspiration (PET)

The PET values estimated by the four methods shared a concurrent spatial pattern and temporal trend. Calcula-

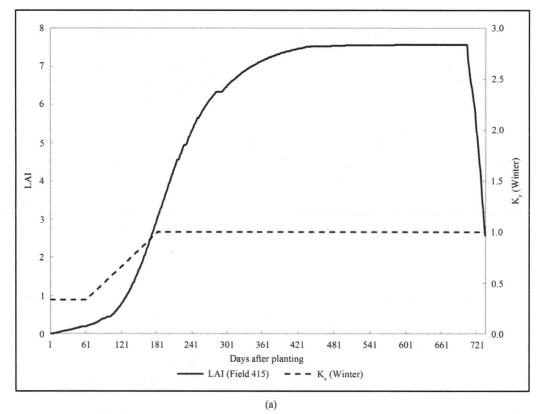

(a)

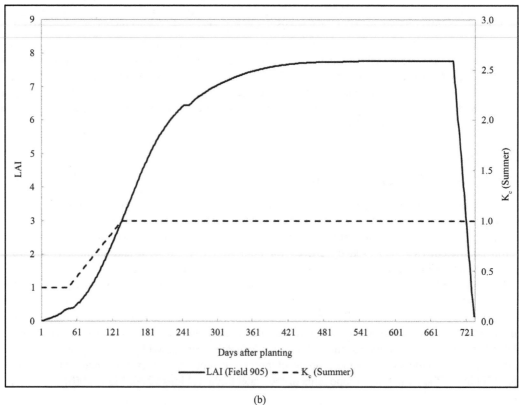

(b)

Figure 3. Leaf area index (LAI) compared to crop coefficient (K_c) for (a) the summer crop on field 415 and (b) the winter crop on field 905.

tion of PET with Hargreaves or Priestley-Taylor produced values that strongly deviate from the HC&S and Penman-Monteith methods. These differences are more obvious during the summer months (May to October) than during the winter months (November to April). A comparison of daily series of PET obtained with different methods is presented in **Figure 4(a)**, **Figure 4(b)**. At Field 415, predicted annual PET is 1964.3 mm using the

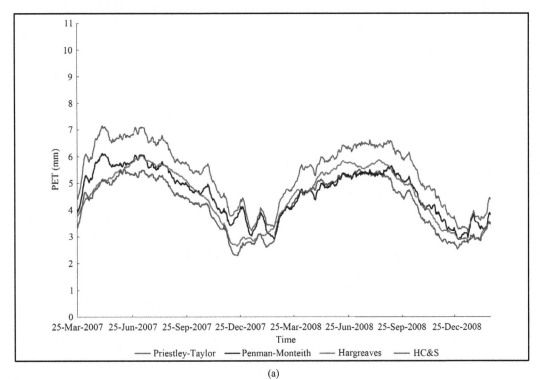

(a)

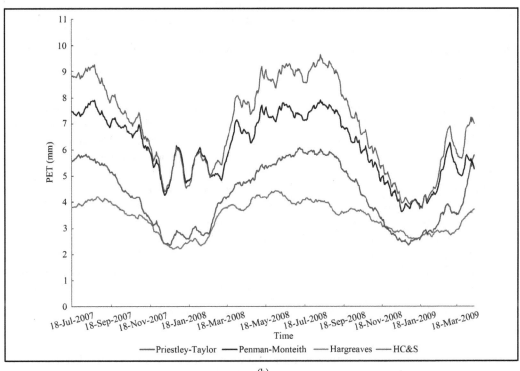

(b)

Figure 4. Comparison of PET calculated by the four tested methods at Fields 415 (a) and 905 (b).

HC&S method, 1693.6 mm using Penman-Monteith, 1656.5 mm using Hargreaves, and 1542.2 mm using Priestley-Taylor. At the nearest site, measured pan evaporation has an average of 2265.5 mm ± 193.3 mm. The relative errors of the simulated values with respect to pan measurements are 13.3% for the HC&S method, 25.2% for Penman-Monteith, 26.9% for Hargreaves, and 31.9% for Priestley-Taylor. At Field 905, PET calculated using the HC&S, Penman-Monteith, Priestley-Taylor, and Hargreaves methods is 2320 mm, 2059.5 mm, 1447.2 mm, and 1157.9 mm, respectively. Measured pan evaporation nearest to Field 905 has an average of 2466.55 mm ± 244.5 mm. The relative errors are 5.9%, 16.5% mm, 41.3%, and 53.1% for the HC&S, Penman-Monteith, Priestley-Taylor, and Hargreaves methods, respectively.

4.3. Actual Evapotranspiration (AET)

SWAT calculates AET based on LAI or the crop coefficient method until LAI equals 3. After that, the crop is assumed to be fully developed and the model assumes that all PET is partitioned to plant water uptake and soil evaporation is assumed to be negligible. However, regardless of the crop stage, soil moisture is the limiting factor that determines the volume of water that is evaporated in a given day. This critical process is well represented in SWAT. **Figure 5** shows the effect of shorter periods of low soil water contents, which reduce the AET rate to the minimum possible. A comparison of AET methods for both fields 415 and Field 905 are presented in **Figure 5(a)**, **Figure 5(b)**. Annual AET values predicted by SWAT range from 1081 mm to 1544 mm.

Table 1(a) and Table 1(b) show the average AET values and the corresponding standard deviations for four different crop stages. The crop cycle for sugarcane was divided into growth stages defined by the break points of the crop coefficient curves used by HC&S for winter and summer plantings. At Field 415 the highest average values of AET were found during the germination and establishment phase for all methods, while at Field 905 the highest average AET values correspond to the grand growth phase. One reason for this behavior is that both fields experienced several days of water stress during the grand growth phase, which caused AET to stop completely.

In addition, the ratio between AET estimated by the Penman-Monteith and HC&S methods were calculated for the period of time where PET is all partitioned to plant uptake and no soil evaporation occurs (grand growth phase). The ratio AET/PET can be interpreted as being equivalent to the pan coefficients, which are ratios of PET to pan evaporation for a given vegetative land cover. The calculated ratios are 1.16 ± 0.11 and 1.14 ± 0.12 for fields 415 and 905, respectively. The ratios found in this study are consistent with reported coefficients for sugarcane fields under Hawaiian conditions. USGS [37] used a sugarcane pan coefficient of 1.18 during the middle stage of growth. Reported by McMahon *et al.* [2], sugarcane can have a pan coefficient as high as 1.2.

4.4. Water Balance

For hydrologic studies, the major purpose of estimating AET is to determine actual water that will be lost from the system (**Figure 6**). The SWAT simulated water balance for the crop period (2007-2009) is shown in **Table 2**. The study site at Field 415 receives a total of 3149 mm (686 mm precipitation + 2463 mm irrigation). Field 905 receives a total of 2079 mm (675 mm precipitation + 2863 mm irrigation). For both fields, precipitation has a very small contribution to the water budget (27.8% and 23.4%). Most of the water supply was provided by irrigation (72.2% and 76.6%). While runoff and lateral flow explain between 2.4% to 5.1% of the water balance, AET and percolation account for 94.9% to 97.6% of the water balance calculated with the HC&S method. According to the four applied methods Priestley-Taylor, Penman-Monteith, Hargreaves, and HC&S, AET accounts for 66%, 71%, 69%, and 76% and for 67%, 79%, 58%, and 81% of the total water supply at Field 415 and Field 905, respectively. The relative contribution of the runoff is very small for both fields (0.4% to 4.3%). Lateral flow also shows little variation among different methods (7.8% - 16.4%). The proportions of percolated water represent 26.5%, 24.4%, 32.1% and 16.9% of the water supply calculated with the Priestley-Taylor, Penman-Monteith, Hargreaves, and HC&S methods respectively.

5. Discussion

Comparison of the results of PET estimates using the four methods facilitated the identification of the most suitable method given a specific level of data availability and considering the specific characteristics of the growing environment. The Hargreaves method, which relies only on air temperature, failed to represent refer-

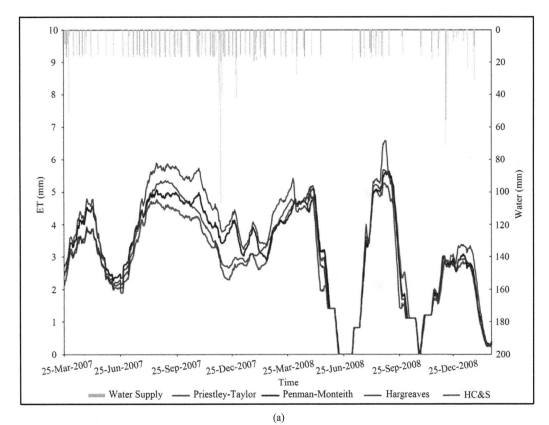

(a)

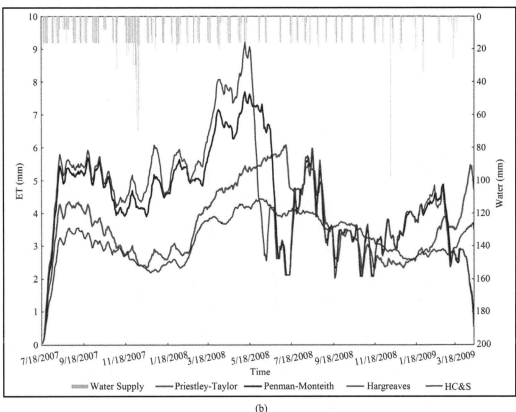

(b)

Figure 5. Comparison of actual evapotranspiration on Fields 415 (a) and 905 (b).

Table 1. Average and standard deviation of evapotranspiration by method and crop stage expressed in mm·day^{-1} for Fields 415 (a) and 905 (b).

(a)

Crop Phase	Hargreaves		Priestley-Taylor		Penman-Monteith		HC&S	
	Mean	SD	Mean	SD	Mean	SD	Mean	SD
Germination	3.6	1.3	4.1	1.5	3.6	1.2	4.3	1.9
Tillering	3.1	1.4	3.4	1.4	3.3	1.5	3.6	1.8
Grand Growth	3.2	1.8	3.5	1.9	3.4	1.9	3.8	2.3
Ripening	2.3	2.0	2.3	2.2	2.3	2.1	2.4	2.5
All Crop Period	3.0	1.8	3.2	2.0	3.1	1.9	3.4	2.4

(b)

Crop Phase	Hargreaves		Priestley-Taylor		Penman-Monteith		HC&S	
	Mean	SD	Mean	SD	Mean	SD	Mean	SD
Germination	3.9	1.4	4.9	1.8	3.1	1.1	5.2	2.0
Tillering	3.1	1.0	4.7	1.4	2.7	0.7	5.1	1.6
Grand Growth	4.5	1.8	5.1	3.0	3.8	0.6	5.1	3.8
Ripening	3.2	1.5	3.2	2.3	3.0	0.6	3.2	2.6
All Crop Period	3.8	1.6	4.4	2.6	3.3	0.8	4.5	3.1

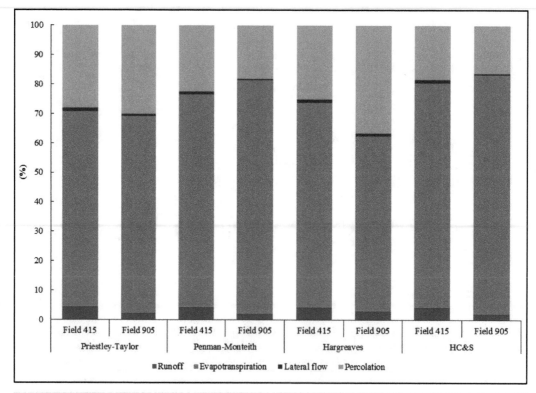

Figure 6. Water balance for Fields 415 and 905 expressed in percent of water supply.

Table 2. Water balance for Fields 415 (a) and 905 (b) expressed in mm of water. Results of the 2-year crop period.

(a)

Component	Priestley-Taylor	Penman-Monteith	Hargreaves	HC&S
Total Water Inflow	**3166.9**	**3162.4**	**3165.1**	**3147.6**
Soil Available Water Content	17.9	13.4	16.1	−1.4
Total Water Supply	3149.0	3149.0	3149.0	3149.0
Precipitation	685.7	685.7	685.7	685.7
Irrigation	2463.2	2463.2	2463.2	2463.2
Total Water Outflow	**3128.5**	**3132.8**	**3130.1**	**3145.0**
Runoff	135.7	131.8	132.2	129.4
Evapotranspiration	2079.5	2264.7	2175.8	2403.4
Lateral Flow	35.7	32.8	34.2	30.8
Percolation	877.7	703.5	787.8	581.5

(b)

Component	Priestley-Taylor	Penman-Monteith	Hargreaves	HC&S
Total Water Inflow	**3567.5**	**3567.9**	**3644.0**	**3567.6**
Soil Available Water Content	0.0	0.4	76.5	0.0
Total Water Supply	3567.6	3567.6	3567.6	3567.6
Precipitation	675.8	675.8	675.8	675.8
Irrigation	2891.7	2891.7	2891.7	2891.7
Total Water Outflow	**3561.5**	**3561.1**	**3485.3**	**3561.2**
Runoff	75.1	67.5	95.8	67.4
Evapotranspiration	2389.7	2828.8	2083.5	2896.9
Lateral Flow	25.0	17.6	28.5	16.4
Percolation	1071.7	647.2	1277.5	580.4

ence evapotranspiration in irrigated regions of Maui under windy conditions. In that regard, Hargreaves and Allen [23] stated that daily estimates by the Hargreaves equation are subject to error because of the influence of the temperature range caused by large variations in wind speed. Both Hargreaves and Priestley-Taylor methods produced predictions that tended to be smaller than those predicted by the Penman-Monteith and HC&S methods. Dingman [8] concluded that temperature is not a reliable indicator of sunlight and generally is a poor predictor of pan evaporation because air temperature in Hawaii is greatly modified by the marine surroundings. Under the conditions of the study site, whenever soil water content is not the limiting factor, differences in wind velocity account for most of the variation in PET calculated with different methods. In general, PET methods, that require wind and especially humidity data, such as Penman-Monteith and HC&S, will simulate windy and humid conditions more accurately [14] [38]. The observed trends of AET as a function of crop growth stage failed to match the trends reported in other publications in which AET rises from germination phase through tillering and reaches the maximum value at the grand growth phase to then decline during the ripening phase. For instance, Freire [39] reported AET values of 3.0 mm·day^{-1} for the germination phase, 3.8 mm·day^{-1} for tillering phase, 5.1 mm·day^{-1} for the grand growth period, and 3.1 mm·day^{-1} for the ripening phase. One reason for this result could be water stress. Sugarcane at both fields had to endure a lack of rainfall and irrigation for several days at late states of the crop development. As reported, AET in sugarcane fields is highly variable largely due to local climatic conditions, hydrologic settings, and the type of production systems. However, predicted AET made by SWAT for the entire crop cycle with Penman-Monteith and HC&S methods in two different field conditions compared well to reported measured data of AET that ranges between 2.33 - 5.70 mm·day^{-1} [39]-[42].

Using the HC&S method, the proportion of percolated water (16.3% - 18.5%) compares reasonable well to 19% and 34% of recharge found as a proportion of the total water input by Izuka *et al.* [43] and Shade [44]. In addition, AET proportions calculated with the HC&S method (76.4% - 81.3%) are also supported by the comparisons made with measured Pan A evaporation datasets. Accordingly, it can be assumed that SWAT properly represents the system under analysis when using the HC&S method to account for AET. HC&S managers use their own modified ET equation to plan irrigation. In both fields (415 and 905), the amounts of irrigation (2463.2 mm and 2896.9 mm), *i.e.* the water supply, and AET (2403.4 mm and 2896.9 mm), *i.e.* the demand, are similar.

6. Conclusions

Considering the important role that evapotranspiration plays in the water cycle, the appropriate selection of the PET method to estimate AET is of great importance. Commonly, the PET method is selected based on the availability of climatic datasets to feed the models. The lack of climatic variables limits the use of physically sound PET approaches. However, the selection of the PET method should rather be taken based on site characteristics; the use of a PET method that takes into account the dominant climatic variables is highly recommended. By ignoring the dominant weather factor, predicted PET can be meaningless and likely not representative of the environmental conditions that need to be assessed. Based on the results of this study, the HC&S method appropriately represents the specific climatic and environmental conditions of the central valley in Maui, in which HC&S sugarcane plantations are located. However because the HC&S method was developed for very specific conditions, the Penman-Monteith method is recommended for other parts of the Hawaiian Islands, where climatic conditions are dominated by wind characteristics. In order to represent local conditions, PET during the grand growth phase of sugarcane should be calculated multiplying the Penman-Monteith PET by a coefficient that ranges between 1.14 and 1.16.

The present study assessed the influences of PET methods on the water balance of sugarcane cropping system in the Hawaiian Island of Maui. The results suggests that the proportions of ET, percolated water, lateral flow and runoff compare reasonably well with results reported in other studies performed on the Island of Maui. Accordingly, the HC&S Company allocates water for irrigation based on an appropriate estimation of AET. However, additional work should be done to analyze the entire system that includes water collection from the ditch system (including bypass water), analysis of ground water recharge and pumping system and the water budget for the entire island.

Acknowledgements

This study was funded by the US Department of Navy, Office of Naval Research & USDA-ARS Bioenergy Project in Hawaii. The authors are grateful to Mae Nakahata from HC&S for providing research access and assistance and to the anonymous reviewers for their valuable suggestions to improve the paper.

References

[1]   Grubert, E. (2011) Freshwater on the Island of Maui: System Interactions, Supply, and Demand. Master's Thesis, Environmental Water Resource Engineering, University of Texas at Austin, Austin.

[2]   McMahon, T.A., Peel, M.C., Lowe, L., Srikanthan, R. and McVicar, T.R. (2012) Estimating Actual, Potential, Reference Crop and Pan Evaporation Using Standard Meteorological Data: A Pragmatic Synthesis. *Hydrology and Earth System Sciences*, 9, 11829-11910. http://dx.doi.org/10.5194/hessd-9-11829-2012

[3]   Lloyd, B.R., Davidson, J.R. and Hogg, H.C. (1972) Estimating the Productivity of Irrigation Water for Sugarcane Production in Hawaii. *Economic Research Service Technical Report No.* 56, USDA—Natural Resource Economics Division, 72.

[4]   National Agricultural Statistics Service (NASS) (2013) National Statistics for Sugarcane. http://www.nass.usda.gov

[5]   Baumgartner, A. and Reichel, E. (1975) The World Water Balance. R. Oldenbourg-Verlag, Munchen, Wien, 1-179.

[6]   Irmak, S., Howell, T.A., Allen, R.G., Payero, J.O. and Martin, D.L. (2005) Standardized ASCE Penman-Monteith: Impact of Sum-Of-Hourly vs. 24-Hour Timestep Computations at Reference Weather Station Sites. *Transactions of the*

ASAE, **48**, 1063-1077. http://dx.doi.org/10.13031/2013.18517

[7] Wang, X., Melesse, A.M. and Yang, W. (2006) Influences of Potential Evapotranspiration Estimation Methods on Swat's Hydrologic Simulation in a Northwestern Minnesota Watershed. *Transactions of the ASABE*, **49**, 1755-1772. http://dx.doi.org/10.13031/2013.22297

[8] Dingman, S.L. (1994) Physical Hydrology. Macmillan College Publishing Company, New York, 575.

[9] Allen, R.G., Pereira, L.S., Raes, D. and Smith, M. (1998) Crop Evapotranspiration: Guidelines for Computing Crop Water Requirements. FAO Irrigation and Drainage Paper 566, UN-FAO, Rome.

[10] McCuen, R.H. (1998) Hydrologic Analysis and Design. 2nd Edition, Prentice-Hall, Upper Saddle River.

[11] Giambelluca, T.W. and Nullet, D. (1992) Evaporation at High Elevations in Hawaii. *Journal of Hydrology*, **136**, 219-235. http://dx.doi.org/10.1016/0022-1694(92)90012-K

[12] Farahani, H.J., Howell, T.A., Shuttleworth, W.J. and Bausch, W.C. (2007) Evapotranspiration: Progress in Measurement and Modeling in Agriculture. *Transaction of the ASABE*, **50**, 1627-1638. http://dx.doi.org/10.13031/2013.23965

[13] Shih, S.F. (1987) Using Crop Yield and Evapotranspiration Relations for Regional Water Requirement Estimation. *Water Resources Bulletin—American Water Resources Association*, **23**, 435-442.

[14] Gracel, B. and Quickl, B. (1998) A Comparison of Methods for the Calculation of Potential Evapotranspiration under the Windy Semi-Arid Conditions of Southern Alberta. *Canadian Research Hydrology*, **13**, 9-19.

[15] Blumenstock, D.I. and Price, S. (1967) Climate of Hawaii. In: *Climates of the States*, *No.* 60-51, *Climatography of the United States*, US Department of Commerce, Washington DC.

[16] Jensen, M.E., Burman, R.D. and Allen, R.G. (1990) Evapotranspiration and Irrigation Water Requirements. American Society of Civil Engineers Manual and Reports on Engineering Practice No. 70, New York, 360.

[17] Bezuidenhout, C.N., Lecler, N.L., Gers, C. and Lyne, P.W.L. (2006) Regional Based Estimates of Water Use for Commercial Sugarcane in South Africa. *Water South Africa*, **32**, 219-222.

[18] Arnold, J.G. and Fohrer, N. (2005) SWAT2000: Current Capabilities and Research Opportunities in Applied Watershed Modeling. *Hydrolocical Processes*, **19**, 563-572. http://dx.doi.org/10.1002/hyp.5611

[19] Gassman, P.W., Reyes, M., Green, C.H. and Arnold, J.G. (2007) The Soil and Water Assessment Tool: Historical Development, Applications, and Future Directions. *Transactions of the ASABE*, **50**, 1211-1250. http://dx.doi.org/10.13031/2013.23637

[20] Green, W.H. and Ampt, G.A. (1911) Studies on Soil Physics, Part 1, the Flow of Air and Water through Soils. *Journal of Agricultural Science*, **4**, 11-24.

[21] Monteith, J.L. (1965) Evaporation and the Environment. 19*th Symposia of the Society for Experimental Biology*, **19**, 205-234.

[22] Sumner, D.M. and Jacobs, J.M. (2005) Utility of Penman-Monteith, Priestley-Taylor, Reference Evapotranspiration, and Pan Evaporation Methods to Estimate Pasture Evapotranspiration. *Journal of Hydrology*, **308**, 81-104. http://dx.doi.org/10.1016/j.jhydrol.2004.10.023

[23] Hargreaves, G.H. (1975) Moisture Availability and Crop Production. *Transactions of the ASAE*, **18**, 980-984. http://dx.doi.org/10.13031/2013.36722

[24] Priestley, C.H.B. and Taylor, R.J. (1972) On the Assessment of Surface Heat Flux and Evaporation Using Large-Scale Parameters. *Monthly Weather Review*, **100**, 81-82. http://dx.doi.org/10.1175/1520-0493(1972)100<0081:OTAOSH>2.3.CO;2

[25] Hargreaves, G.H. and Allen, R.G. (2003) History and Evaluation of Hargreaves Evapotranspiration Equation. *Journal of Irrigation and Drainage Engineering*, **129**, 53-63. http://dx.doi.org/10.1061/(ASCE)0733-9437(2003)129:1(53)

[26] Ekern, P.C. and Chang, J.H. (1985) Pan Evaporation: State of Hawaii, 1894-1983. Hawaii. Department of Land and Natural Resources, Division of Water and Land Development, Honolulu, Rep. R74, viii + 172.

[27] Sanderson, M. (1993) Prevailing Trade Winds: Weather and Climate in Hawaii. University of Hawaii Press, Honolulu, 126.

[28] Mink, J.F. and Lau, L.S. (2006) Hydrology of the Hawaiian Islands. University of Hawaii Press, Honolulu.

[29] NOAA (National Oceanic and Atmospheric Administration) (2013) Comparative Climatic Data for the United States through 2012. http://ols.nndc.noaa.gov/plolstore/plsql/olstore.prodspecific?prodnum=C00095-PUB-A0001

[30] Anderson, R.G. and Wang, D. (2014) Energy Budget Closure Observed in Paired Eddy Covariance Towers with Increased and Continuous Daily Turbulence. *Agricultural and Forest Meteorology*, **184**, 204-209. http://dx.doi.org/10.1016/j.agrformet.2013.09.012

[31] Gesch, D.B. (2007) The National Elevation Dataset. In: Maune, D., Ed., *Digital Elevation Model Technologies and Ap-*

plications: *The DEM User's Manual*, 2nd Edition, American Society for Photogrammetry and Remote Sensing, Bethesda, 99-118.

[32] Fry, J.A., Coan, M.J., Homer, C.G., Meyer, D.K. and Wickham, J.D. (2009) Completion of the National Land Cover Database (NLCD) 1992-2001 Land Cover Change Retrofit Product. *US Geological Survey Open-File Report* 2008-1379, 18.

[33] Nair, S.S., King, K.W., Witter, J.D., Sohngen, B.L. and Fausey, N.R. (2011) Importance of Crop Yield in Calibrating Watershed Water Quality Simulation Tools. *JAWRA*: *Journal of the American Water Resource Association*, **47**, 1285-1297. http://dx.doi.org/10.1111/j.1752-1688.2011.00570.x

[34] Jones, C.A. (1980) A Review of Evapotranspiration Studies in Irrigated Sugarcane in Hawaii. *Hawaiian Planters' Record*, **59**, 195-214.

[35] Liu, W.Z., Hunsaker, D.J., Li, Y.S., Xie, X.Q. and Wall, G.W. (2002) Interrelations of Yield, Evapotranspiration, and Water Use Efficiency from Marginal Analysis of Water Production Functions. *Agricultural Water Management*, **56**, 143-151. http://dx.doi.org/10.1016/S0378-3774(02)00011-2

[36] Allison, F.E., Roller, E.M. and Raney, W.A. (1985) Relationship between Evapotranspiration and Yields of Crops Grown in Lysimeters Receiving Natural Rainfall. *Agronomy Journal*, **50**, 506-511. http://dx.doi.org/10.2134/agronj1958.00021962005000090004x

[37] United States Geological Service (USGS) (2007) Effects of Agricultural Land-Use Changes and Rainfall on Ground-Water Recharge in Central and West Maui, Hawaii, 1926-2004. *Scientific Investigations Report* 2007-5103, 69.

[38] Suleiman, A.A. and Hoogenboom, G. (2007) Comparison of Priestley-Taylor and FAO-56 Penman-Monteith for Daily Reference Evapotranspiration Estimation in Georgia. *Journal of Irrigation and Drainage Engineering*, **133**, 175-182. http://dx.doi.org/10.1061/(ASCE)0733-9437(2007)133:2(175)

[39] Da Silva, T.G.F.F. (2009) Análise de crescimento, interação biosfera-atmosfera e eficiência do uso de água da cana-de-açúcar irrigada no submédio do vale do São Francisco. TeseDoutorado—Universidade Federal de Viçosa, 194.

[40] Souza, E.F., Bernardo, S. and Carvalho, J.A. (1999) Função de produção da cana—Deaçúcaremrelação à água para três cultivares, em Campos dos Goytacazes. *Engenharia Agicola*, **19**, 28-42.

[41] Inman-Bamber, N.G. and McGlinchey, M.G. (2003) Crop Coefficients and Water-Use Estimates for Sugarcane Based on Long-Term Bowen Ratio Energy Balance Measurements. *Field Crops Research*, **83**, 125-138. http://dx.doi.org/10.1016/S0378-4290(03)00069-8

[42] Santos, M.A.L. (2005) Irrigação suplementar da cana-de-açúcar (Saccharumspp.): Um modelo de análise de decisão para o Estado de Alagoas. Tese Doutorado—Escola Superior de Agricultura "Luiz de Queiroz", Piracicaba, 101.

[43] Izuka, S.K., Oki, D.S. and Chen, C. (2005) Effects of Irrigation and Rainfall Reduction on Ground-Water Recharge in the Lihue Basin, Kauai, Hawaii. *Scientific Investigations Report* 2005-5146, US Department of the Interior, 48.

[44] Shade, P.J. (1999) Water Budget of East Maui, Hawaii. *Water-Resources Investigations Report* 98-4159, US Department of the Interior, 36.

7

Recharge to Blue Lake and Strategies for Water Security Planning, Mount Gambier, South Australia

Nara Somaratne[1]*, Jeff Lawson[2], Glyn Ashman[1], Kien Nguyen[3]

[1]South Australian Water Corporation, Adelaide, Australia
[2]Department of Environment, Water and Natural Resources, Mount Gambier, Australia
[3]Hydraulic Works and Management Division, Directorate of Water Resources, Ministry of Agriculture and Rural Development, Hanoi, Vietnam
Email: *nara.somaratne@sawater.com.au

Abstract

Blue Lake, a volcanic crater provides municipal water supply to the city of Mount Gambier, population of 26,000. Current average annual pumping from the lake is 3.6×10^6 m^3. The lake is fed by karstic unconfined Gambier Limestone aquifer. Storm water of the city discharges to the aquifer via about 400 drainage wells and three large sinkholes. Average annual storm water discharge is estimated at approximately 6.6×10^6 m^3 through drainage wells and sinkholes within 16.8 km^2 of the central part of the city. Chemical mass balance for calcium was used to estimate groundwater inflow to the lake at 6.3×10^6 m^3, almost equal to the volume of storm water discharge and slightly higher than the previous estimates using environmental isotopes ($4.8 - 6.0 \times 10^6$ m^3). Considering the lake outflow volume of 2.7×10^6 m^3, the net inflow to the lake equates to the current annual pumping and therefore it is considered that the current pumping rate is at the upper limit. For meeting the short-term future demand, confined aquifer water may be used and in the longer-term, an additional well field is required outside the Blue Lake capture zone, preferably to the north-east of the city. For water supply security, inflow to the lake along with water quality has to be maintained within the city. Current annual private abstraction within the capture zone is about 4.4×10^6 m^3 and in order to maintain aquifer water levels, no additional allocation should be allowed.

Keywords

Blue Lake, Recharge, Water Security Planning, Groundwater, Water Resource Management

---

*Corresponding author.

1. Introduction

Blue Lake, a picturesque volcanic crater, is the source of drinking water supply to the city of Mount Gambier, a large regional city of South Australia, population about 26,000. Additionally, Blue Lake is an important tourist attraction due to its annual colour change cycle, which is driven by calcite precipitation [1]-[4]. The lake is groundwater-fed through an extensive karst aquifer [5] [6].

The water balance of this groundwater-fed lake has been altered through urbanisation since settlement in the mid-to-late 1800s [7]. As a result, the average water residence time within the lake declined from 23 ± 2 years to 8 ± 2 years by the late twentieth century [8]. Urbanisation has produced a threefold increase in deposition of calcite in Blue Lake [8], and it is possible that the annual calcite precipitation cycle provides a mechanism for determining annual inflow to the lake [9]. The average depth of the lake is approximately 72 m and the surface area is 6.1×10^5 m^2. Owing to the steep slopes of the crater, the surface water catchment area is limited to 8.6×10^5 m^2, only slightly greater than the surface area of the lake itself [3] [7]. A water balance analysis of the lake confirms that the main input is groundwater, of which a considerable portion is storm water recharge. Current annual extraction from Blue Lake for drinking water supply is approximately 3.6×10^6 m^3, with a peak annual extraction of 4.2×10^6 m^3 during the 1970s (**Figure 1**).

Estimation of recharge to Blue Lake is critical to water supply security. Few studies compare hydrological and chemical estimates of seepage to and from lakes. The net contribution of groundwater to lakes is commonly estimated as the difference between measured gains and losses of water from stream flow, precipitation, and evaporation [10] [11]. Another approach to determining the groundwater component of lake water balances is to use chemical mass balance of major ions [12].

A comprehensive evaluation via the flow net approach is possible, where groundwater seepage is a major component of water and chemical balances [13] [14]. For example, a study of Williams Lake, Minnesota, focused on water and chemical balances from 1980 through 1991 [14].

To estimate recharge to Blue Lake, Turner [15] used environmental isotopes and concludes that 80% of annual inflow to the lake results in leakage from the confined aquifer. Ramamurthy [9] used environmental isotopes ^{234}U and ^{238}U and interpretation of hydrochemical data of the aquifer and lake; in this study, the basic strategy was to compute mass balance calculations of U and Ca in the lake and interpret the results from its recharge regime. Turner *et al.* [6] re-examined inflow to the lake using environmental isotopes ^{3}H, ^{2}H, ^{18}O and ^{14}C in lake water and groundwater. Similar to Ramamurthy [9], Turner *et al.* [6] assumed that only 10% of inflow to the lake comes from the confined aquifer through upward leakage. Herczeg *et al.* [8] studied alteration in catchment subsurface water balance through land use changes on the lake; and changes to the water and carbon budget of the lake using sediment isotope records ($\delta^{18}O_{carb}$, $\delta^{13}C_{carb}$, and $\delta^{13}C_{org}$).

We examined the water balance of the lake using mass balance of the calcium. Investigation includes water sampling and analysis for major ion chemistry of the aquifer, analysis of available sediment core samples for major ion chemistry, and recharge estimation of the capture zone immediately up-gradient to the lake.

2. Study Area

Mount Gambier is the main city in the Lower South East of South Australia (**Figure 2**), a region dependent

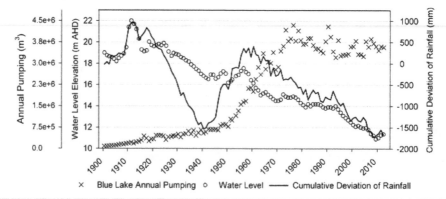

Figure 1. Blue Lake water level, annual pumping and cumulative deviation of mean annual rainfall.

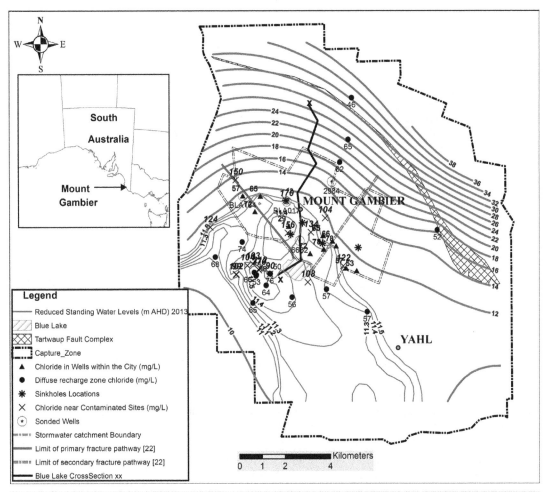

Figure 2. Location map of the Blue Lake and capture zone.

primarily on groundwater for water supply [16]. Since the 1970s, the city has experienced steady growth in population with resultant development activity [17]. Blue Lake is the largest of three lakes formed in a volcanic complex in the Lower South East of South Australia [18]. As is typical of many crater lakes, Blue Lake is steep-sloped, has a very small surface catchment, and is relatively deep for its surface area [1] [6]. Blue Lake has a warm monomictic stratification regime (*i.e.*, mixes once a year; Wetzel [19]), is oligotrophic (*i.e.*, has a low biological productivity), and algal production is probably phosphorus limited [2]. The surrounding region has a temperate climate, with annual mean maximum temperature of 18°C and mean minimum of 8°C. Average annual rainfall is 750 mm and pan evaporation is about 1400 mm.

Hydrogeological Setting

The study area is located within the Gambier Basin, a mixed sequence of marine and terrestrial deposits [20]. The main geological units in the Mount Gambier area, in downward order, consist of Holocene volcanic deposits, the Bridgewater Formation (stranded Pleistocene beach dunes), the Gambier Limestone and the Dilwyn Formation [21]. The unconfined aquifer within the Gambier Limestone is a continuous system and an important supplier of groundwater throughout the region. Karstic features are common within the Gambier Limestone. With the exception of local areas where direct disposal of organic wastes occurs, the aquifer is generally well oxygenated and with relatively low salinity (300 - 600 mg·L^{-1}). Underlying the Gambier Limestone is the Dilwyn Formation, which comprises a series of unconsolidated sands with carbonaceous clay interbeds (**Figure 3**). The Dilwyn Formation hosts a confined aquifer with higher salinity than the Gambier Limestone. A major fault underlies the Blue Lake, which is thought to represent a zone of regional structural weakness through which

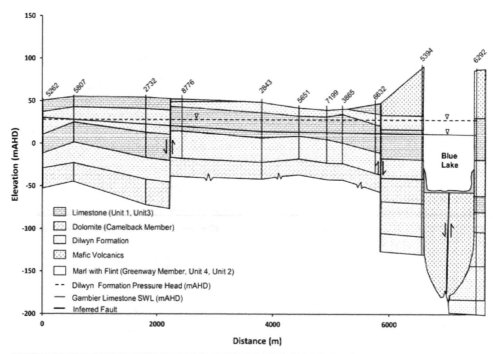

Figure 3. North-south cross-section across the city-XX in **Figure 2**.

volcanic activity has occurred.

Recently, Lawson [22] refined the stratigraphy of the Gambier Limestone surrounding the Blue Lake, defining the subunits within the Gambier Limestone in hydrostratigraphic terms (**Figure 3**). The following are general descriptions of the unconfined aquifer hydrostratigraphic unit characteristics as described by Li *et al.* [23].

Green Point Member Unit 1 (Unit 1) is a transmissive bryozoal limestone with marl and flint inclusions. Green Point Member Unit 2 (Unit 2) is composed of marls and flints, has the potential to act as a semi-confining unit and is characterised by generally lower transmissivities with associated low hydraulic conductivity. Green Point Member Unit 3 (Unit 3) is described as an open transmissive bryozoal limestone. It contains minor fracturing and may be partly dolomitised. This unit is characterised by higher aquifer transmissivities and associated hydraulic conductivity. Green Point Member Unit 4 (Unit 4) is typically described as a deep marl section within the limestone. It may act as an aquitard with typically associated lower hydraulic conductivity. Camelback Member is a dual porosity unit, characterised by extensive fracturing at the subunit intersection point. This results in extremely high porosities and hydraulic conductivities. The non-fractured sections of the subunit are composed of fine dolomitic rhombs and have transmissivities probably similar to or lower than bryozoal sections of limestone. Greenways Member is typically described as grey basal marl with resultant low transmissivities and hydraulic conductivity (**Figure 3**).

3. Methods

3.1. Field Measurements

Existing groundwater data from the capture zone and the Blue Lake were used in this study. Additional groundwater samples for major ion chemistry analysis were taken from the Blue Lake capture zone, within and outside the city, to supplement existing knowledge and to fill data gaps. Groundwater samples for major ion chemistry analysis were collected using the micro-purge (low-flow) sampling procedure [24] along with grab sampling. Micro sampling is employed to gain representative groundwater samples within the open-hole sections of monitoring and drainage wells. Low-flow purging is considered [24] superior to bailing and high-rate pumping and results in a more representative sample than the typical three-volume, well-purge methodology. The assumption when using grab sampling is that the hydrostratigraphy in the well is in hydraulic equilibrium prior to sampling. To collect the sample by this method, an electronic depth sampler connected to a geophysical logging line is advanced to the target sampling depth and the unit is electronically opened, allowing groundwater to enter the sampler. After a short wait, the sampler is closed again. Salinity profiles of monitoring and drainage wells were obtained using Hydrolabsonde [25] connected to a logging truck cable and lowered down the well from the surface to the well base, recording electrical conductivity (EC) data along the way.

3.2. Recharge to the Capture Zone

Gambier Limestone water level contours indicate a potential for ground water flow toward Blue Lake from all points within the lake's ground water capture zone (**Figure 2**). Aquifer parameters derived from pumping tests are limited in number and show large variation. The water table fluctuation method for recharge assessment is therefore discounted as a suitable method. Long-term average recharge to the capture zone may be obtained using the conventional chloride mass balance method, but of most interest to this study is recharge in that part of the capture zone, where point recharge through about 400 drainage wells [26] and 3 sinkholes directly recharges the aquifer system in the city. The area selected for this purpose is the inflow area between the primary fracture pathway and secondary fracture pathway (**Figure 2**) of Lawson [22]. Recharge to the inflow zone of Blue Lake is assessed using the generalized chloride mass balance method described in [27]:

$$R = \frac{\left(Pc_{p+D}\right)+Q_p\left(c_{gd}-c_s\right)}{c_{gd}} \tag{1}$$

where R (LT^{-1}) is the recharge, P (LT^{-1}) is average annual rainfall, c_{p+D} (ML^{-3}) is chloride concentration of rainfall including contribution from dry deposition, c_s (ML^{-3}) is chloride concentration of surface water, Q_p (LT^{-1}) is runoff to sinkholes (point recharge source) and c_{gd} (ML^{-3}) is groundwater chloride concentration in the diffuse recharge zone. Note that Q_p is expressed as depth of catchment. In the absence of direct measurement, c_{p+D} can be estimated from [28] using:

$$c_{p+D} = 35.45 \times \left\{\frac{0.99}{d^{0.25}} - 0.23\right\} \tag{2}$$

where "d" is distance in kilometers from the ocean in the prevailing wind direction and c_s is taken from surface water sampling for chloride analysis. An important input parameter for Equation (1) is Q_p, which is taken from Nguyen [29]. For brevity, a brief description of the modelling procedure is provided below.

As part of the recharge estimation for Blue Lake in Mount Gambier, Nguyen [29] used the urban storm water model MUSIC [30] for quantifying storm water runoff to drainage wells. In that study, rainfall and runoff processes were modelled for the period 2007-2012 using a daily time-step with daily rainfall and evaporation data. For sub-catchments with drainage wells, average percentage of impervious (51%) and pervious (49%) areas were determined from digital maps of the city using geographic information system tools. A rainfall threshold of 1 mm was used for impervious areas. Uniform soil storage capacity and field capacity values of 120 mm and 80 mm were used for the pervious areas. Initial soil storage capacity was set to 30%. Average annual runoff volume from both pervious and impervious areas were calculated as point recharge to drainage wells and sinkholes. A sensitivity analysis indicates field capacity of the soil had the greatest effect on runoff from the previous area [29]. Storm water derived from the central 16.8 km^2 of the city area (26.5 km^2) is discharged to the unconfined aquifer through three sinkholes and about 400 storm water drainage wells. For the average annual rainfall of 750 mm, 6.6×10^6 m^3 of runoff volume flows through drainage wells and sinkholes to groundwater. Out of total runoff volume, about 5.4×10^6 m^3 is generated from the impervious areas of the catchment.

3.3. Inflow to Blue Lake

Ramamurthy [9] argues that when the Blue Lake unconfined aquifer water level is compared to the confined aquifer water level, a positive head difference of about 20 m is observed, creating the possibility of upward leakage through a volcanic conduit, if present. An outcome of this scenario would be resultant elevated chloride values in water chemistry sampled from the lake's base. Conventional through-flow calculation using Darcy's equation or flow net approach [14] is not followed in this study due to the extreme heterogeneity of the Gambier Limestone aquifer. The transmissivity of the aquifer is in the range of 450 - 24,000 $m^2 \cdot day^{-1}$ and specific yield is 0.1 - 0.4 for the Gambier Limestone [31]. As an alternative approach, water and chemical mass balance is used to estimate ground water inflow and outflow for lakes [32]. This is in contrast to the typical water budget approach, which allows only the net ground water flow to be computed. In addition, when using steady-state assumptions, combined with the water budget, the chemical mass balance approach computes long-term estimates of ground water inflow and outflow [32]. The approach used in this study is to compute ground water inflow-outflow using a mass balance calculation of Ca in the lake and to interpret the results with respect to the groundwater flow regime.

Although Na and Cl are conservative elements, large variation in recorded values in the aquifer due to contaminated sites preclude mass balance calculations. In addition to the contribution from contaminated sites, Ramamurthy [9] suggests leaching of Na and Cl ions from basaltic rocks is a likely additional source. As a result, chloride is not used for the chemical mass balance calculation for the lake; this is computed using mass balance of the Ca ion in this study.

Higher Ca^{2+} concentrations in the surrounding groundwater were observed than those recorded in the lake. This is due to the reaction of calcite with carbon dioxide derived from oxidation of organic matter in the aquifer. The fundamental basis for this is that carbon dioxide reacts with water to form carbonic acid (H_2CO_3), providing protons (H^+), which associate with the carbonate ion (CO_3^{2-}) from calcite to form bicarbonate (HCO_3^-). The overall reaction between CO_2 and $CaCO_3$ is [33]:

$$CO_2 + H_2O = H_2CO_3 \tag{3}$$

$$H_2CO_3 + CaCO_3 = Ca^{2+} + 2HCO_3^- \tag{4}$$

This reaction is fundamental to understanding the behaviour of $CaCO_3$ dissolution and precipitation in nature. An increase in CO_2 results in dissolution of $CaCO_3$. Removal of CO_2 causes $CaCO_3$ to precipitate. This degassing of CO_2 in the lake causes $CaCO_3$ to precipitate, therefore reducing Ca^{2+} in the lake water. Another process of removal of CO_2 in lakes is photosynthesis, which consumes CO_2. Therefore, with knowledge of Ca^{2+} in the aquifer, the lake and its removal through pumping and sedimentation, it is possible to calculate inflow to the lake using the water and Ca^{2+} budgets.

A water budget for the lake can be written as [8]:

$$\Delta V / \Delta t = I + R - E - O - P \tag{5}$$

where ΔV is average annual change in storage (L^3), Δt is time (T), I is annual inflow (L^3T^{-1}) to the lake, O is annual outflow volume from the lake (L^3T^{-1}), P is annual extraction volume (L^3T^{-1}), R is annual volume of rainfall to the lake (L^3T^{-1}) and E is annual evaporation (L^3T^{-1}).

The general equation for the chemical budget of the lake is [14]:

$$\nabla(Vc_l)/\Delta t = Ic_{gwi} - Oc_l - Pc_l - S \tag{6}$$

where S is the amount of ion deposited in the lake sediment (MT^{-1}), c is chemical concentration (ML^{-3}), subscript l is for lake and subscript "gwi" is for groundwater inflow. Equation (6) assumes any change in c due to R and E is negligible.

4. Results and Discussion

4.1. Water Balance

Water samples for analysis were taken from monitoring wells where stratigraphy had been interpreted. Water chemistry data presented in **Table 1** are from the capture zone surrounding Blue Lake and include wells located near historically known contaminated sites. Ramamurthy [9] noted that anomalously high concentrations of Na, Cl and Uranium toward the northwest of the lake were likely caused by leaching of basaltic rocks. Average chloride data indicate consistent distribution across unconfined aquifer subunits, with the value in the Camelback Member markedly lower than the recorded concentrations in the Dilwyn Formation confined aquifer. This indicates that there is no significant upward leakage to the Gambier Limestone aquifer from the confined aquifer despite the pressure head difference being positive. Slightly elevated calcium ion in aquifer Unit 4 is an indication of low hydraulic conductivity of the subunit, where water flow is slow, thus increasing the contact time with limestone.

The large variation in EC may be due to non-uniform mixing with sub-aquifer waters through leakage, diffuse

Table 1. Water quality parameters of the unconfined aquifer. Sample size, n, is given in bracket [after 22].

Water quality parameter	Gambier Limestone aquifer sub-unit				Overall unconfined aquifer (n = 33)	Dilwyn confined aquifer (n = 8)
	Unit 1 (n = 11)	Unit 3 (n = 9)	Unit 4 (n = 4)	Camelback (n = 9)		
Ca (mg·L^{-1})	58.3 ± 11.4	64.2 ± 19.8	68.5 ± 25.9	56.4 ± 12.9	61 ± 16.3	73.5 ± 7.8
Mg (mg·L^{-1})	28.9 ± 9.4	32.6 ± 5.1	24.4 ± 16.6	24.5 ± 14.4	28.1 ± 11.1	29.7 ± 2
K (mg·L^{-1})	13.7 ± 7.4	13.3 ± 9.6	5.9 ± 2.7	9 ± 10.3	11.4 ± 8.7	7.6 ± 1
Na (mg·L^{-1})	73.9 ± 24.9	75.3 ± 25.9	67.3 ± 19.7	59.4 ± 18.7	69.5 ± 23	111.9 ± 12.7
Cl (mg·L^{-1})	83.9 ± 38.1	85 ± 43.5	83.3 ± 19.5	84.8 ± 29.6	84.3 ± 34.1	160.5 ± 16
SO$_4$ (mg·L^{-1})	32.6 ± 30.1	21.1 ± 11.5	20.9 ± 18.3	14 ± 5.1	23.1 ± 20.7	18.9 ± 1.8
HCO$_3$ (mg·L^{-1})	208.9 ± 72	256.4 ± 79.6	217.8 ± 117.7	242.2 ± 89	231.3 ± 83.6	334.8 ± 62.6
Free CO$_2$ (mg·L^{-1})	17.1 ± 13.5	21.9 ± 7.6	11.5 ± 8.7	17.7 ± 9.8	17.8 ± 10.7	-
EC (μS·cm^{-1})	848 ± 243.3	899.4 ± 293.6	843 ± 310	697 ± 221.5	820.5 ± 260	11650 ± 62.6
pH	7.4 ± 0.2	7.2 ± 0.2	7.6 ± 0.2	7.4 ± 0.2	7.4 ± 0.2	7.4 ± 0.1

recharge and point recharge via drainage wells and sinkholes. An interesting point to note is that the lowest EC is found in the deepest sub-aquifer, the Camelback Member, even though the drainage wells are open-hole construction below the water table and generally extend to this subunit. This confirms that the Camelback Member is the major pathway carrying recharge water from drainage wells to the lake. Chloride data obtained in this study is comparable with the Ramamurthy study [9], where an average chloride value of 81.3 ± 10.4 (n = 10) is reported. Chloride values obtained from monitoring wells away from contaminated sites (63 ± 26, n = 16) are the same as chloride values obtained in the diffuse recharge zone outside the city boundary (**Figure 2**), 62.9 ± 9 (n = 13), indicating that point recharge through drainage wells has not altered the chloride in the point recharge zone.

Blue Lake water chemistry data are presented against the stratigraphy of the north wall of the lake (**Figure 4**). An increased calcium value (**Figure 4(a)**) at Unit 4 level corresponds to the enhanced Ca ion in the aquifer. Significantly decreased Ca ion in the lake is due to CaCO$_3$ precipitation as CO$_2$ degassing takes place and is reflected in the reduced free CO$_2$ profile in the lake. Other than Ca, HCO$_3^-$ and free CO$_2$, variations of all other parameters in the lake are insignificant, indicating that the lake is well-mixed.

Even though the chloride values within both the lake and outside groundwater are similar (**Table 1** and **Figure 4**), about 82 to 85 mg·L^{-1}, the chloride value in the Dilwyn confined aquifer is significantly higher at around 160 mg/L (160.5 ± 16 mg·L^{-1}). This indicates that upward leakage from the confined aquifer through the volcanicfault is insignificant, and thus an assumption can be made that the lake water is solely derived from the unconfined limestone aquifer.

Groundwater inflow to the lake in this study is based on mass balance Equations (1) and (2). For the chemical mass balance, Ca^{2+} is taken, because no excessive variation in Ca^{2+} ion occurs in the aquifer when compared to Na or Cl ions (**Table 1**). The annual rate of recharge to the lake is calculated on the basis of the calcium ion in the aquifer, the lake and the sedimentation rate in the lake. The mass balance calculation for calcium was done on the basis of a single mean value taken from the aquifer, and the lake. This approach envisages that groundwater inflow to the lake is exclusively from the unconfined aquifer. The amount of Ca precipitated in the lake (**Table 2**) is calculated for two depositional periods, historical and contemporary. As is evident in **Table 2**, the sediment accumulation rate has increased in recent years, indicating higher precipitated calcite as CO$_2$ is degassed, implying more groundwater enters the lake.

Herczeg *et al.* [8] report that the volumetric porosity of sediment cores range between 0.8350 and 0.875 throughout most cores, resulting in a volumetric dry bulk density of 0.125 - 0.165 g·cm^{-3}. Porosity increases to 0.910 over the top 25 mm, with a corresponding decrease in bulk density to 0.090 g·cm^{-3} at the top of the core. With the assumption of dry bulk density of 2.5 g·cm^{-3}, Herczeg *et al.* [8] estimate the chronological age of individual sediment layers up to a depth of 130 mm by ^{210}Pb-derived mass accumulation rate estimates. Accordingly,

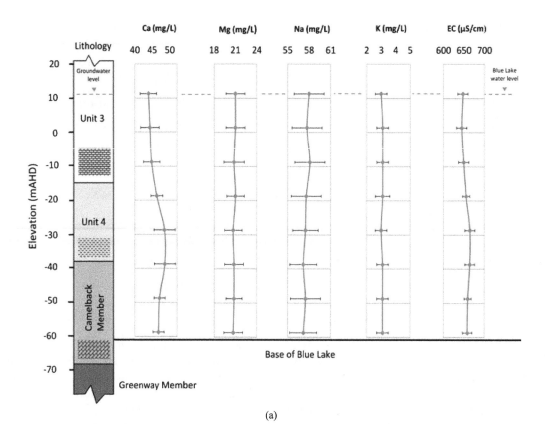

(a)

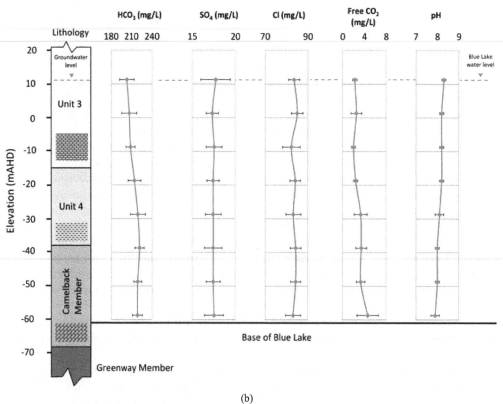

(b)

Figure 4. Water quality data for Blue Lake (data source: South Australian Environmental Protection Authority, unpublished data). (a) Cations and EC; (b) Anions and free CO_2 and pH.

140 mm depth of sediment corresponds approximately with the year 1915. Using the sediment depth versus the year of the deposition relation of [8], it is estimated that the high accumulation rate began in the 1940-1950 period (78 mm depth), which corresponds to acceleration of pumping, as shown in **Figure 1**. The 130 mm depth corresponds to approximately a 1920 date of deposition, and therefore, the pumping rate in 1920 (337×10^3 $m^3 \cdot yr^{-1}$) was taken as the historical pumping rate for the calculation.

The input parameters and calculated inflow to and outflow from the lake are given in **Table 3**. Based on calcium precipitation, contemporary inflow and outflow calculated in this study are higher than the Herczeg et al. [8] estimate but comparable to the results of [9] and [6]. Ramamurthy [9] calculate the lower limit of annual recharge to the lake at the present time as 4×10^6 $m^3 \cdot yr^{-1}$ with an upper limit of 6.6×10^6 $m^3 \cdot yr^{-1}$. Turner et al. [6] use the environmental isotopes ^3H, ^2H, ^{18}O and ^{14}C in the lake and groundwater to estimate the rate of groundwater through-flow in the lake. The water balance calculations of [6] indicate a total groundwater inflow to Blue Lake of between 5.0 and 6.5×10^6 $m^3 \cdot yr^{-1}$, which is in good agreement with the Ramamurthy (1983) estimate. Both [9] and [6] consider that 90% of groundwater inflow to the lake comes from the unconfined aquifer, and ignore the possibility of groundwater outflow from the lake.

Based on isotope mass balance of the sediment ($\delta^{18}O_{carb}$), Herczeg et al. [8] determine the historical inflow to the lake at 2.8×10^6 m^3 and outflow at 2.48×10^6 m^3 from the lake, and contemporary inflow of 4.8×10^6 m^3 with outflow of 0.73×10^6 m^3. Total inflow to the lake calculated using calcite precipitation in the lake is slightly higher than the average total inflow of 6.0×10^6 m^3 determined by Ramamurthy [9] and Turner et al. [6]. The net inflow to the lake from calcite precipitation (3.6×10^6 m^3) corresponds to current annual pumping and is an indication that the pumping rate has reached its upper limit. Recharge to the aquifer is calculated using equation (1) for the inflow area (14 km^2), which corresponds to an area between the primary and secondary fracture

Table 2. Blue Lake sediment analysis for major ions.

| Sediment depth (mm) | Accumulation rate [mg/(cm²·yr)] | | | | | | | Bq·kg⁻¹ |
	Sediment	Ca	Mg	Na	K	Cl	SO₄	²¹⁰Pb
10	106	33.7	0.76	0.04	0.03	0.009	0.03	99.2
20	92	29.8	0.65	0.04	0.02	0.008	0.02	98.4
32	88	28.7	0.6	0.03	0.02	0.004	0.02	75.7
45	87	29.7	0.55	0.03	0.02	0.004	0.005	58.3
60	105	33	0.86	0.04	0.05	0.007	0.007	35.7
78	120	35.8	1.02	0.05	0.09	0.012	0.012	36.2
91	55	14.9	0.47	0.03	0.06	0.007	0.03	20.2
130	52	12.7	0.46	0.03	0.07	0.006	0.03	15.2

Table 3. Input parameters and inflow and outflow to blue lake.

Parameter	Historical (1910-1920)	Contemporary average (1975-1995)	Units	Reference
Pumping	0	3.6	$m^3 \cdot yr^{-1} \times 10^6$	
Rainfall	0.434	0.434	$m^3 \cdot yr^{-1} \times 10^6$	Herczeg et al. (2003)
Evaporation	0.724	0.724	$m^3 \cdot yr^{-1} \times 10^6$	Herczeg et al. (2003)
Calcium deposition	12.7	30.5	$mg \cdot cm^{-2} \cdot yr^{-1}$	
c_{gwi}	61	61	$mg \cdot L^{-1}$	
c_l	43.7	43.7	$mg \cdot L^{-1}$	
Inflow	3.3	6.3	$m^3 \cdot yr^{-1} \times 10^6$	
Outflow	3.0	2.7	$m^3 \cdot yr^{-1} \times 10^6$	

flow paths [22], within the city limits with parameters for Equation (1) taken as: P (750 mm·yr^{-1}), c_{p+D} (14.3 mg·L^{-1}), c_{gd} (69 mg·L^{-1}), Q_p (390 mm·yr^{-1}) and c_s (14 mg·L^{-1}). This results in an average annual recharge volume of 6.85×10^6 m^3, which is comparable with present day annual inflow to the lake. An interesting point to note is that contemporary inflow to the lake is satisfied by point recharge through drainage wells.

Slightly higher water levels in aquifer Unit 1 in the southern face (**Figure 2**) also suggest groundwater inflow to the lake as originally reported in [4]. In this study, inflow from subunit 1 is assumed to be minor and is not included in the water balance calculation. In contrast, lower groundwater levels in aquifer subunits 2 and 3 on the southern side suggest these units are the major outflow from the lake [22]. Turner *et al.* [6] suggest that the groundwater release zone of flow through lakes can be delineated by mapping the distribution of enriched stable isotope compositions of lake water in the surrounding aquifer. Identification of aquifer subunits that carry outflow from the lake was not carried out, as currently there are insufficient data from the southern side of the lake; this is a subject left for future investigation.

4.2. Management Strategies for Water Security

Determination of a water balance is an important factor in the management of water supplies. This paper indicates that the current level of extraction is at the upper limit of a sustainable yield and that an additional water source is required to satisfy increases in demand. To satisfy an increase in demand, a planning strategy is required that identifies both potential threats and constraints to the water supply and opportunities for additional supply sources. The current town water supply to the city of Mount Gambier is reliant on Blue Lake, with two wells constructed in the confined aquifer as emergency water supply.

Two factors need to be considered for water supply security of the lake. First, maintaining current pumping levels from the unconfined aquifer, and second, protection of water quality. In the absence of surface water supplies in the area, ground water supplies from the unconfined aquifer are used for irrigation. Currently, a total of 4.4×10^6 m^3 is extracted for private irrigation use from the capture zone. While there is no significant extraction that directly intercepts flow paths to Blue Lake, an increase in extraction from the south and southeast would affect the flow regime and hence the lake's water balance. Thus, limiting extraction to current levels is suggested, with no additional allocation permitted within the capture zone.

In a comparison of historical water quality data for Blue Lake and for groundwater in the unconfined Gambier Limestone aquifer with water quality target values, Vanderzalm *et al.* [34] [35] show no potential for breach of these water quality guideline values. Trace metal and metalloid data illustrate several historical peaks in concentration, but no evidence suggests rising concentrations within Blue Lake, since co-precipitation occurs with the lake's calcite precipitation process [35].

Based on average population growth in the city 2001-2006 period of 0.8%, a population increase of up to 33,000 could occur by 2030 [17]. With current extraction from the lake already at its upper limit, alternative water sources need to be developed to satisfy increasing demand. For short-term demand increases, the confined aquifer water is the most likely source. The limitations of this option are the need for the wells to be sited to ensure cost effective chlorination and blending, and the ratio of confined to unconfined water that can be achieved while maintaining acceptable hardness and salinity levels.

For long-term planning, an additional unconfined aquifer wellfield is required targeting the Camelback Member subunit. Location of such a wellfield will be the subject of additional investigations. Establishing a wellfield on the southern side of the lake is dependent on the capacity of the water distribution network. A wellfield on the north side beyond the Tartwarp fault complex is preferred, as the city is expanding toward the north. In the northwestern part of the city, the target subunit Camelback Member occurs at a shallow depth [22], and hence is prone to contamination. Therefore, the northeastern side beyond the capture zone boundary may be a potentially suitable area for investigation for a future wellfield.

5. Conclusion

A water balance calculation based on calcite precipitation in Blue Lake provides comparable results with previous water balance studies based on environmental isotopes. Current net inflow to the lake equates to the current extraction level, and recharge through drainage wells equates to the total inflow to the lake. This indicates extraction from the lake has reached its upper limit, and recharge via drainage wells forms an important component of the lake's water balance. While short-term demand increase can be satisfied from the confined aquifer, an alternative wellfield in the unconfined aquifer is recommended to meet long-term water demand.

Acknowledgements

The authors thank Joanne Vanderzalm for providing Blue Lake sediment core samples for major ion chemistry analyses and discussion on sediment deposition periods. Daryl Morgan of Mount Gambier City Council is thanked for providing catchment data for the rainfall-runoff model. The editor and reviewers are thanked for evaluating the manuscript.

References

[1] Tamuly, T. (1969) Physical and Chemical Limnology of the Blue Lake of Mount Gambier. School of Physical Sciences, Flinders University, Adelaide, Research Paper No. 28.

[2] Allison, G.B. and Harvey, P.D. (1983) Freshwater Lakes, in Natural History of South East. In: Tyler, M.J., Twidale, C.R.T., Ling, J.K. and Holmes, J.W., Eds., Royal Society of South Australia, 61-74.

[3] Telfer, A.L. (2000) Identification of Processes Regulating the Colour and Colour Change in an Oligotrophic, Groundwater-Fed Lake, Blue Lake, Mount Gambier, South Australia.

[4] Turoczy, N.J. (2002) Calcium Chemistry of Blue Lake, Mount Gambier, Australia, and Relevance to Remarkable Seasonal Colour Changes. Archiv fur Hydrobiologie, 156, 1-9. http://dx.doi.org/10.1127/0003-9136/2002/0156-0001

[5] Waterhouse, J.D. (1977) The Hydrogeology of the Mount Gambier Area. Department of Mines, Geological Survey of South Australia, Report of Investigations 48.

[6] Turner, J.V., Allison, G.B. and Holmes, J.W. (1983) Environmental Isotope Methods for the Determination of Lake-Groundwater Relations: Applications to Determine the Effect of Man's Activities. International Conference on Groundwater and Man, Sydney, 1.

[7] Lamontagne, S. and Herczeg, A. (2002) Predicted Trends for NO3-Concentration in the Blue Lake, South Australia, Consultancy Report for South Australian Environment Protection Authority, CSIRO Land and Water.

[8] Herczeg, A.L., Leaney, F.W., Dighton, J.C., Lamontagne, S., Schiff, S.L., Telfer, A.L. and English, M.C. (2003) A Modern Isotope Record of Changes in Water and Carbon Budgets in a Groundwater-Fed Lake: Blue Lake, South Australia. Limnology and Oceanography, 48, 2093-2105. http://dx.doi.org/10.4319/lo.2003.48.6.2093

[9] Ramamurthy, L.M. (1983) Environmental Isotope and Hydrogeochemical Studies of Selected Catchments in South Australia. Ph.D. Thesis, School of Earth Sciences, Flinders University, Adelaide.

[10] Likens, G.E. (1985) An Ecosystem Approach to Aquatic Ecology—Mirror Land and Its Environment. Springer-Verlag, New York. http://dx.doi.org/10.1007/978-1-4613-8557-8

[11] Staubitz, W.W. and Zariello, P.J. (1989) Hydrology of Two Headwater Lakes in the Adirondack Mountains of New York. Canadian Journal of Fisheries and Aquatic Sciences, 46, 268-276. http://dx.doi.org/10.1139/f89-037

[12] Srauffer, R.E. (1985) Use of Solute Tracers Released by Weathering to Estimate Groundwater Inflow in Seepage Lakes. Environmental Science Technology, 19, 405-411. http://dx.doi.org/10.1021/es00135a003

[13] Brown, B.E. and Cherkauer, D.S. (1992) Phosphate and Carbonate Mass Balances and Their Relationships to Groundwater Inputs at Beaver Lake, Waukesha County, Wisconsin. Wisconsin Water Research Center.

[14] LaBaugh, J.W., Winter, T.C., Rosenberry, D.O., Schuster, P.F., Reddy, M.M. and Aiken, G.R. (1997) Hydrological and Chemical Estimates of the Water Balance of a Closed-Basin Lake in North-Central Minnesota. Water Resources Research, 33, 2799-2812. http://dx.doi.org/10.1029/97WR02427

[15] Turner, J.V. (1979) The Hydrologic Regime of Blue Lake, South Australia. Ph.D. Thesis, School of Earth Sciences, Flinders University, Adelaide.

[16] Emmet, A.J. (1985) Mount Gambier Storm Water Quality. Department of Engineering and Water Supply Report 84/23.

[17] Planning, S.A. (2008) Greater Mount Gambier Master Plan. Government of South Australia, Strategic and Social Planning Division, Planning SA.

[18] Sheard, M.J. (1983) Volcanoes. In: Tyler, M.J., Twidale, C.R.T., Ling, J.K. and Holmes, J.W., Eds., Natural History of Southeast, Royal Society of South Australia, Adelaide, 7-14.

[19] Wetzel, R.G. (1983) Limnology. 2nd Edition, Saunders College Publishing, Philadelphia.

[20] Drexel, J.F. and Preiss, W.V. (1995) The Geology of South Australia, Vol. 2, The Phanerozoic.

[21] Love, A.J. (1991) Groundwater Flow Systems: Past and Present, Gambier Embayment, Otway Basin, South-East Australia. Ph.D. Thesis, Flinders University of South Australia, Adelaide.

[22] Lawson, J.S. (2013) Water Quality and Movement of the Unconfined and Confined Aquifers in the Capture Zone of

the Blue Lake, Mount Gambier, South Australia and Implications for Their Management. Master's Thesis (Unpublished), The University of South Australia, South Australia.

[23] Li, Q., McGowran, B. and White, M.R. (2000) Sequences and Biofacies Packages in the Mid-Cenozoic Gam-Bier Limestone, South Australia: Reappraisal of Foraminiferal Evidence. *Australian Journal of Earth Sciences*, **47**, 955-970. http://dx.doi.org/10.1046/j.1440-0952.2000.00824.x

[24] Vail, J. (2011) Groundwater Sampling, U.S. Environmental Protection Agency, Science and Ecosystem Support Division, Athens, Georgia.

[25] Eco Environmental (2013) Salinity Sonde, Environmental Monitoring and Sampling Equipment. http://ecoenvironmental.com.au

[26] EPA (2007) EPA Guidelines for Stormwater Management in Mount Gambier. Environment Protection Authority, South Australia.

[27] Somaratne, N. (2012) Pitfalls in the Application of the Chloride Mass Balance Method to Groundwater Basins Dominated by Point Recharge. SA Water Hydrogeological Research Report, SA Water/2012/1 (Unpublished).

[28] Hutton, J.T. (1976) Chloride in Rainwater in Relation to Distance from Ocean. *Search*, **7**, 207-208.

[29] Nguyen, K. (2013) Estimating the Annual Storm Water Yield in the Blue Lake Capture Zone, South Australia. M. Eng thesis, Unpublished, University of South Australia, South Australia.

[30] MUSIC Development Team (2009) Model for Urban Stormwater Improvement Conceptualization (MUSIC). eWater CRC.

[31] Mustafa, S. and Lawson, J.S. (2002) Review of Tertiary Aquifer Properties, Gambier Limestone, Lower Southeast, South Australia. Department of Water, Land and Biodiversity Conservation, Report DWLBC 2002/24, Adelaide.

[32] Sacks, L.A., Herman, J.S., Konikow, L.F. and Vela, A.L. (1992) Seasonal Dynamics of Groundwater-Lake Interactions of Donana National Park, Spain. *Journal of Hydrology*, **136**, 123-154. http://dx.doi.org/10.1016/0022-1694(92)90008-J

[33] Appelo, C.A.J. and Postma, D. (2007) Geochemistry, Groundwater and Pollution. 2nd Edition, A.A. Balkema Publishers, Leiden.

[34] Vanderzalm, J.L., Dillon, P.J., Page, D., Marvanek, S., Lamontagne, S., Cook, P., King, H., Dighton, J., Sherman, B. and Adams, L. (2009) Protecting the Blue Lake from Land Use Impacts. CSIRO Water for a Healthy Country National Research Flagship.

[35] Vanderzalm, J.L., Dillon, P.J., Hancock, G.J., Leslie, C., Dighton, J., Smith, C. and Pearce, G. (2013) Using Elemental Profiles in the Sediment of a Lake Used to Supply Drinking Water to Understand the Impacts of Urban Storm Water Recharge. *Marine and Fresh Water Research*, CSIRO Publishing. http://www.publish.csiro.au/?paper=MF12215

8

Prediction of Ground Water Level in Arid Environment Using a Non-Deterministic Model

Mohammad Mirzavand[1], Seyed Javad Sadatinejad[2], Hoda Ghasemieh[1], Rasool Imani[1], Mehdi Soleymani Motlagh[1]

[1]Department of Natural Resources and Geoscience, University of Kashan, Kashan, Iran
[2]Department of New Sciences and Technologies, University of Tehran, Tehran, Iran
Email: mmirzavand23@yahoo.com

Abstract

Modeling and forecasting of the groundwater water table are a major component of effective planning and management of water resources. One way to predict the groundwater level is analysis using a non-deterministic model. This study assessed the performance of such models in predicting the groundwater level at Kashan aquifer. Data from 36 piezometer wells in Kashan aquifer for 1999 to 2010 were used. The desired statistical interval was divided into two parts and statistics for 1990 to 2004 were used for modeling and statistics from 2005 to 2010 were used for valediction of the model. The Akaike criterion and correlation coefficients were used to determine the accuracy of the prediction models. The results indicated that the AR(2) model more accurately predicted ground water level in the plains; using this model, the groundwater water table was predicted for up to 60 mo.

Keywords

Non-Deterministic Models, Akaike Criterion, Ground Water Level, Kashan Aquifer

1. Introduction

The conventional method for predicting hydrological variables is to use time series analysis. The first attempts to study time series, particularly in the nineteenth century, recognized the benefits of economic measures [1]. This was developed by Yule (1927), who examined different time series to detect random processes. Since then, types of time series have expanded. Hydrological time series consists of a time-dependent hydrological variable, such as the flow rate of a river. The purpose of hydrological studies is to quantitatively describe the statistical population; the process of creating this statistical population was based on a limited number of samples [2]. Random time series have been applied to solving hydrological issues by Brass and Rodriguez (1985) [3]; Berkol and Davis (1987) [4], and Lin and Lee (1994) [5]. Mathematical modeling of time series can produce hydrological synthetic data, predict hydrological events, identify trends and shifts in data, complete a missing data, and generate data. The output of artificial series from river flows for drought and flood studies have been used to optimize utilization of reservoir systems and design the capacity for water supply systems [2] [6], among other uses. One basic hypothesis for modeling time series is that time series is stationary. A time series with its own statistical parameters (e.g., mean and variance) is considered stationary when its expected value is time-independent. The expected value is the average value which would be expected if the processes were repeated an infinite number of times. More formally, the expected value is a weighted average of all possible values. Many hydrological time series are non-stationary for reasons such as climate change, drought, and statistical parameters such as mean and standard deviation. To increase knowledge about methods of statistical determination, it could be useful to remove stationary and non-stationary time series [7]. Factors that may cause non-static time series include periodic or seasonal trends and shifts [8]. Static tests can detect the impact of each factor on a stationary time series; however, non-stationary examinations of time series can aid understanding of the physical mechanisms, which indicate the impotence of static tests on hydrological time series analysis (Wang *et al.*, 2005) [9].

Stationary time series analysis methods generally fall into two categories. The first category consists of methods based on the analysis if there is a statistical difference in segments of a time series. In the second category, the static test is based on the statistical properties of the whole time series [10]. Numerous studies have been done in the field of hydrology based on time series, such as Javidi and Sharifi (2009), who used time series to predict the mean annual flow rate of a river. Evaluation of time series models using the Akaike information criterion (AIC) and residual variance has concluded that the autoregressive (2) (AR (2)) model is more appropriate for data production, thus, it was chosen as the final model [11]. Khalili *et al.* (2010) investigated the trends and stationary analysis of river flow in Urumia province for Shahr Chay River using KPSS and ADF methods. Their results showed that the annual flow rate series was static at a significance level of 5% [7]. Golmohammadi and Safavi (2010) used univariate hydrological time series and a fuzzy system based on adaptive neural networks to predict the flow rate of Zayandehrood, the largest river in the central plateau of Iran. The results indicated the efficiency of these systems in the forecast [12]. Nakhaei and Mirarabi (2010) used the Box-Jenkins model to predict floods using series data to determine the flow rate for Sumar in Kermanshah province of Iran. Box-Jenkins applies the autoregressive moving average (ARMA or ARIMA) models to finding the best fit of a time series to past values in order to make forecasts. They used trial-and-error criteria of residual and selected the best model among the models examined (ARMA (1, 1, 0) (2, 2, 1) (1, 2)) to predict river flow rate for the next 24 months [13]. Ahan (2000) predicted water table fluctuations using ARIMA models. According to the data, he used the quadratic difference method to remove the trend in time series [14]. Saeidian and Ebadi (2004) determined the time series model for the flow rate of Talkhehrood river in northwestern Iran. They fitted time series patterns for over 53 years of flow rate data to the AIC test and determined that the best model was the AR(2) [15]. Jalali (2002) provided decision support systems (DSS) for repositories of time series models to forecast the input of monthly flow rate for Jiroft Damin Kerman Province using a single-univariate ARIMA model to calibrate the input flow rate [16]. Alonekyenak (2007) examined the performance of artificial neural network models and the ARMA model to predict the level of water for a time period of more than 1 mo [17]. Amabyl *et al.* (2008) fitted time series models to simulate data obtained from the SWAT model and historical data. They found that the groundwater table data performed well in time series models [18]. The present study used the results of previous research and the ability of time series techniques for time series analysis to determine the optimal model and used it to predict the water table of an aquifer and determine the accuracy of predictions.

2. Methods

2.1. Study Area Choice

The study area (longitude: 51°32' to 51°03'E, latitude: 33°27' to 34°13'N) is located in Kashan plain, Esfahan province, Iran (**Figure 1**).

This area is bordered to the north and northwest by Qom province, to the southwest by the Mutah mountain

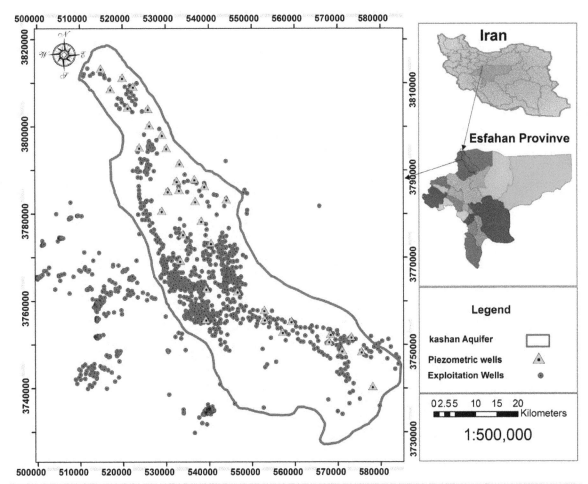

Figure 1. Position of study area and exploitation of Piezometric wells.

range to the south by the Natanz plains and to the east by the Salt Lake range. It has an area of approximately 7083 km², of which 3040 km² consists of highlands and 4043 km² consists of submontane regions to the north and northeast. The area under study is the Kashan aquifer located in the Kashan plain, which is 1570.23 km² in size (**Figure 1**). The annual evaporation range is 2100 to 3000 mm and the average of annual humidity is about 42 percent. Maximumtemperaturesin the summer, the warmest month (July) to 48°C and a minimum temperature of −5°C inthewintertocome. Annual rainfall is varied in salt lake with 75 mm at the east to 300 mm in the southwest mountains. The Kashan aquifer in unconfined aquifer and because of largely discharge from this aquifer for agriculture, industry and drinking, The Kashan aquifer experiences an average annual loss of about 0.57 m and a critical negative budget (about −32 million m³ annual discharge).

In this study, water table data from the aquifer for 1990 to 2004 was used to predict the ground water for the year's 2005 to 2010 using time series and R software environment. The first step for the time series was to chart analysis of the time series data to determine the presence or absence of trends in the data. The second step was to determine trends in the time series and removal it for create stationary data. After examining the static data, the appropriate model was fitted to the data to determine the best model for prediction. If the data showed seasonal trends, data differencing was performed to bring the mean to zero and remove the seasonal trends. The third step was to investigate the normality of the predicted data for model selection. The Kolmogorov-Smirnov test was chosen to evaluate data normality. The models examined in this study were AR, MA, ARMA, ARIMA and SARIMA. It was not necessary to obtain a fitted linear equation or remove trends, because the models performed this. ARIMA and SARIMA used data differencing, so the seasonal data mode was also eliminated. This study first plotted the time series diagram for seasonal and residual (random) trends (**Figure 2**). Since an examination of time series data could be modeled on random data, random data was obtained from a time series for modeling.

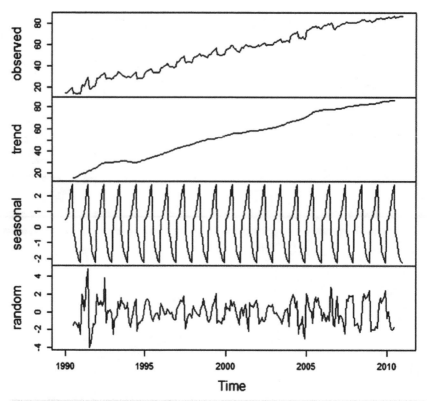

Figure 2. Time series diagram with random, seasonal and trend components.

2.2. Auto-Regressive Model

Researchers such as Slutsky (1937), Walker (1931), Yaglom (1955), and Yule (1927) first formulated the concept of Auto-Regressive (AR) and moving average (MA) [1]. These models are stochastic conventional models and as their name implies, imposes regression on its sentences; however regression is performed for past values of z_t. This model is applicable for stationary and non-stationary time series and the basic structure is suggested in Equation (1):

$$z_t = \phi_1 z_{t-1} + \phi_2 z_{t-2} + \cdots + \phi_p z_{t-p} + a_t \tag{1}$$

In the above equation, ϕ_1, ϕ_2, ϕ_p and ... are coefficients and parameters of the AR model and a_t is random. The remaining time-independent (noise) obeys a normal distribution with a mean of zero. In this model, if $\sum_{j=0}^{\infty} \phi_j$ converges, the process will stationary [19].

2.3. Moving Average Model

The general form of the moving average (MA) model with q rank can be expressed as Equation (2):

$$z_t = \theta_1 a_{t-1} + \theta_2 a_{t-2} + \cdots + \theta_q a_{t-q} + a_t \tag{2}$$

where θ_1, θ_2, θ_q are the coefficients and parameters of the MA model [11].

2.4. General Structure of ARMA Model

The ARMA (p,q) model was created by combining the AR (p) model and the MA (q) model. The general structure of ARMA can be expressed as Equation (3):

$$z_t = \phi_1 z_{t-1} + \phi_2 z_{t-2} + \cdots + \phi_p z_{t-p} + a_t - \theta_1 a_{t-1} - \theta_2 a_{t-2} - \theta_q a_{t-q} \tag{3}$$

where the parameters are the same as for the AR and MA models [11]. Most of the time series are non-stationary in reality; therefore, we can model time series with the static subtraction operation and then an AR or MA model or a complex pattern can be fitted to the differencing series. The result for the non-differential series is a comprehensive model (ARMIA) [20].

2.5. Box-Jenkins Seasonal Pattern ARIMA Model

Non-seasonal time series models in the interconnected autoregressive model are moving averages and are displayed as ARIMA (p,d,q). In this model, p represents the autoregressive model, q represents the moving average model, and d represents data differencing. For a stationary time series, $d = 0$ and the ARIMA model converts to ARMA. The general multiplicative seasonal pattern can be expressed as Equation (4):

$$\phi_p(B)\Phi_p(B^{12})W_t = \theta_q(B)\Theta_Q(B^{12})a_t \qquad (4)$$

where B is the shift operator, and ϕ_p, Φ_p, θ_q, Θ_Q are the polynomials of p, p, q, respectively, and a_t is used instead of z_t (Box-Jenkins) and is a purely random process with a mean of zero and σ_a^2 variance. Variables of the initial set to eliminate trends and seasonality can be expressed as Equation (5):

$$W_t = \nabla^d \nabla_{12}^D X_t \qquad (5)$$

In the ARIMA model, where data differencing occurs d represents non-seasonal differencing. With non-seasonal differencing (D), the ARIMA model is converted to a SARIMA model. The ARIMA time series is the most complete is commonly used. A more detailed discussion about its usage can be found in Box and Jenkins (1976) [19]. Studies by Vangyr and Zoor (1997), Knotters and Vanvalsom (1997) [20], and Ahan (2000) [14] suggest that the Box-Jenkins model for time series is an appropriate model for investigating the behavior of groundwater.

2.6. Determining the Best Model

The properties of the autocorrelation coefficients and partial autocorrelation are other criteria that determine the best model for the time series using AIC. The test is used for comparison of different ARMA (p,q) models and is calculated as Equation (6):

$$AIC(p,q) = N\ln(\sigma_\varepsilon^2) + (p,q) \qquad (6)$$

erehw N is the amount of time series data and σ_ε^2 is the variance of error (residuals). The model with lower AIC is the more appropriate model [21] [22]. In addition to testing for AIC, the length of the forecast for the statistical mean and correlation test are used. The results are shown in **Table 1**. After choosing the appropriate model, the predicted data are normalized and examined using the Kolmogorov-Smirnov model. The data used in this study were monthly; delays for these models were for 12 mo.

3. Results & Discussion

The results of determining a trend in the data and deterministic and random terms (**Figure 2**) to determine autocorrelation and partial autocorrelation functions before removal and data differencing (**Figure 3**). The results for different models are shown in **Table 1**. Models that fit the time series to forecast groundwater level changes are shown in **Figure 4**. **Figure 5** valuates the correlation between the observed and predicted values by the selected models. **Figure 5** shows autocorrelation and partial autocorrelation functions for predicted data by the selected models.

In this study, 5 models using 11 time series structures were examined. For each parameter, AIC and correlation coefficients for the coming months were obtained (**Table 1**).

R software was used to predict the depth of groundwater from 2005 to 2010 using monthly underground water depth data from 1999 to 2004. The important point in is to survey stationary time series of hydrological before modeling of time series, as a result, the static analysis of hydrologic series can be effective in the recognition and interpretation of hydrological processes and its relation to trend and climate change. As we know, the time series data have 4 components (trend component, a seasonal component, jump (maybe there isn't) and a random component) (**Figure 3**). For analyze of the time series data, it is important that deterministic components from the time series be removed. So, after the 3 deterministic components removed, the randomness component used for

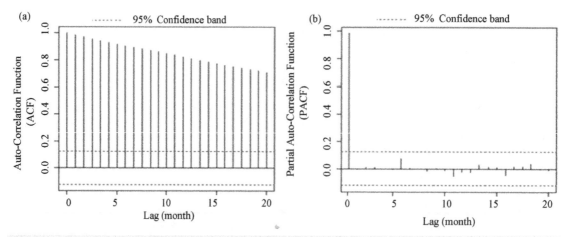

Figure 3. (a) Auto-correlation and (b) partial auto-correlation functions of the monthly groundwater level time series before removal of trends and data differencing.

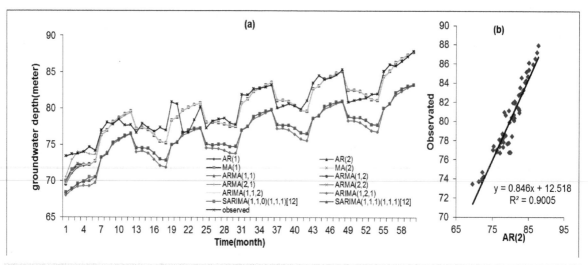

Figure 4. (a) Models prediction versus observed and final model correlation (b).

Table 1. AIC coefficient, parameters and different models used to choose the final model.

Parameters Model Models	$\emptyset 1$	$\emptyset 2$	Θ_1	Θ_2	d	D	S$\emptyset$1	Sθ1	AIC	R2
AR (1)	0.4476	***	***	***	**	**	***	***	522.17	0.8992
AR (2)	0.3905	−0.1036	***	***	**	**	***	***	522.44	0.9005
MA (1)	***	***	0.4164	***	**	**	***	***	524.62	0.9025
MA (2)	***	***	0.4931	0.1998	**	**	***	***	521.66	0.8997
ARMA (1, 1)	0.3174	***	0.1633	***	**	**	***	***	522.93	0.9001
ARMA (1, 2)	−0.0905	***	0.5787	0.2409	**	**	***	***	523.57	0.8998
ARMA (2, 1)	1.3529	−0.5591	−1.000	***	**	**	***	***	502.31	0.909
ARMA (2, 2)	−0.1001	−0.2121	0.5767	0.4290	**	**	***	***	524.84	0.9007
ARIMA (1, 1, 2)	0.3330	***	−0.8453	***	1	**	***	***	524.92	0.8999
ARIMA (1, 2, 1)	−0.2038	***	−1.000	***	2	**	***	***	566.39	0.8884
SARIMA (1, 1, 0) (1, 1, 1) [12]	−0.2196	***	***	***	1	1	−0.0628	−0.9335	553.86	0.8832
SARIMA (1, 1, 1) (2, 2, 1) [12]	0.4294	***	−1.000	***	1	2	−0.0038	−0.9994	520.43	0.8832

modeling. In this study, 5 models with 11 different structures are examined. According to **Figure 3**, obtained before removing deterministic components of the data; it was found that the groundwater depth data has a seasonal trend and it was used to eliminate the trend and create the stationary series used various models. As seen from **Figure 4(a)**, **Figure 4(b)**, the autocorrelation function is exponentially reduced and also, partial autocorrelation function is not significant after a lag. According to the interpretation of these two functions can say at first sight that by using these data the AR model is an appropriate model to predict, but for a more accurate prediction, the Akaike criterion and the correlation coefficient were used to select the final model. The results of various models survey based on test and the Akaike criterion (AIC), which is one of the common methods to compare different models especially ARMA time series, also based on amount of model parameters and correlation coefficients were determined. Between mentioned models, appropriate model is firstly that an amount of the absolute value does not to exceed of 1 and the second it has lowest Akaike criteria (**Table 1**).

Thus, among 11 studying structure based on Akaike criterion and model parameters, the model ARMA (2,1) due to violation of model parameters of 1 were deleted between studying model. Among the remain models; SARIMA model (1, 1, 1) (1, 1, 1) [12] had the lowest Akaike criterion but to examine the results of Akaike criterion for this model (as mentioned in above, Akaike criterion is to test the ARMA models) correlation test was used. As shown in **Table 1**, based on Akaike test SARIMA (1, 1, 0) (1, 1, 1) [12] model has less Akaike statistic value but based on the results of the correlation test had less accuracy of predictions in compared other models. For this reason, another model which had less Akaike and in terms of the correlation coefficient had meaningful predictions was selected. Therefore AR (2) model was selected as the final model, to predict changes in water table levels of Kashan plain for 60 months ahead .The Kolmogorov-Smirnov test was used to examine the normality of the predicated data. The significance was greater than 0.5 (sig = 1), allowed acceptance of the null hypothesis, meaning the normality of predicted data. **Figure 5** shows that the autocorrelation function and partial autocorrelation with 10 time lags after removing the trend data were significant. From the results of studies by Karamooz and Iraqi (2005) [8], Hashemi and Jahanshahi (2005) [22], Javidi and Sharifi (2009) [11], and Nakhaei and Mirarabi (2010) [13] and the results of this study, it can be concluded that the model selected to predicted the function is important and the output is very good. AIC results for all models except the Box-Jenkins model, gave appropriate statistics. It is better use the correlation coefficient to test the models for accuracy of their predictions for the time interval. According to the non-deterministic and randomness nature of hydrological issues, time series are an appropriate technique to forecast hydrological phenomena. According to the results of this research on time series models, for cases were sufficient data is not available, to generate data and predict changes in hydrological variables, time series models can be used. These models can be used to optimize models, for accurate predictions and proper planning, and to obtain reliable management for any area.

4. Conclusion

In this study, the prediction models that were developed with standard logistic regression analysis in groundwater

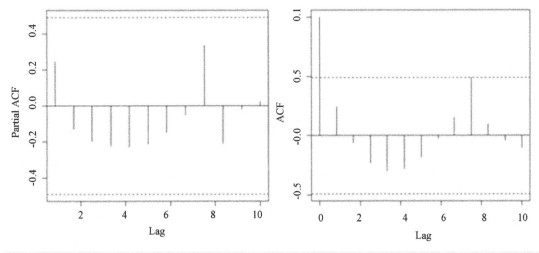

Figure 5. ACF and PACF after removing data.

level data were compared. According to the non-deterministic and random nature of the hydrological issues, time series is one of the appropriate methods to forecast hydrological phenomena.

References

[1]   De Gooijer, J.G. and Hyndman, R.J. (2006) 25 Years of Time Series Forecasting. *International Journal of Forecasting*, **22**, 443-473. http://dx.doi.org/10.1016/j.ijforecast.2006.01.001

[2]   Salas, J.D. (1993) Analysis and Modeling of Hydrological Time Series. In: Maidment, D.R., Ed., *Handbook of Hydrology*, McGraw-Hill, New York, 19.1-19.72.

[3]   Brass, R.L. and Rodriguez-lturbe, L. (1985) Random Functions and Hydrology. Addison-Wesley Publishing Company, Reading.

[4]   Brockwell, P.J. and Davis, R.A. (1987) Time Series: Theory and Methods. Springer-Verlag, New York. http://dx.doi.org/10.1007/978-1-4899-0004-3

[5]   Lin, G.F. and Lee, F.C. (1994) Assessment of Aggregated Hydrologic Time Series Modeling. *Journal of Hydrology*, **156**, 447-458.

[6]   Salas, J.D., Delleur, J.W., Yevjevich, V. and Lane, W.L. (1980) Applied Modeling of Hydrologic Time Series. Water Resources Publications, Littleton.

[7]   Khalili, K., Fakheri Fard, A., Din Pajooh, Y. and Ghorbani, M.A. (2010) Trend and Stationary Analyses of River Flow for Hydrological Time Series Modeling. *Journal of Soil and Water Silences*, **20**, 61-72.

[8]   Karamooz, M. and Aragi Nejad, S. H. (2005) Advanced Hydrology. 1st Edition, Amirkabir University of Technology Press, Tehran, 464.

[9]   Wang, W., Van Gelder, P.H.A.J.M. and Vrijling, J.K. (2005) Trend and Stationary Analysis for Streamflow Processes of Rivers in Western Europe in 20th Century. *IWA International Conference on Water Economics, Statistics and Finance*, Rethymno, 8-10 July 2005, 11.

[10] Chen, H.L. and Rao, A.R. (2003) Linearity Analysis on Stationarity Segments of Hydrologic Time Series. *Journal of Hydrology*, **277**, 89-99. http://dx.doi.org/10.1016/S0022-1694(03)00086-6

[11] Javidi Sabaghian, R. and Sharifi, M.B. (2009) Using Stochastic Models to Simulate River Flow and Forecast Annual Average Flow of the River by Time Series Analysis. *First International Conference on Water Resources*, Semnan, 16-18 August 2009, 9.

[12] Golmohammadi, M.H. and Safavi, H.R. (2010) Pridicting of One-Variable Hydrological Time Series Using Fuzzy Systems Based on Adaptive Neural Network. *The 5th National Congress on Civil Engineering*, Mashhad, 4-6 May 2010, 8 p.

[13] Nakhaei, M. and Mir Arabi, A. (2010) Flood Forecasting through Sumbar River Discharge Time Series Using the Box —Jenkins Model. *Journal of Engineering Geology*, **4**, 901-915.

[14] Ahan, H. (2000) Modeling of Groundwater Heads Based on Second Order Difference Time Series Modeling. *Journal of Hydrology*, **234**, 82-94. http://dx.doi.org/10.1016/S0022-1694(00)00242-0

[15] Saidian, Y. and Ebadi, H. (2004) Time Series Model Determining for Flow Discharge Data (Case Study: Vaniar Hydrometer Station in AjiChay River Basin). *2nd National Student Conference on Water and Soil Resources*, Shiraz, 12-13 May 2004, 7.

[16] Jalali, K. (2002) Jiroft Dam Reservoir Inflow Forecasting Using Time Series Theory. *6th International Seminar on River Engineering*, Ahvaz, 8-10 February 2002, 9.

[17] Alonekyenak, A. (2007) Forecasting Surface Water Level Fluctuations of Lake Van by Artificial Neural Networks. *Water Resource Manage*, **21**, 399-408. http://dx.doi.org/10.1007/s11269-006-9022-6

[18] Amabyl, V., Gabriel, G. and Bernard, A.E. (2008) Fitting of Time Series Models to Forecast Stream Flow and Groundwater Using Simulated Data from SWAT. *Journal of Hydrology Engineering*, **13**, 554-562. http://dx.doi.org/10.1061/(ASCE)1084-0699(2008)13:7(554)

[19] Box, G.E.P. and Jenkins, G.M. (1976) Time Series Analysis: Forecasting and Control. Holden Day, San Francisco.

[20] Knotters, M. and Van Walsum, P.E. (1997) Estimating Fluctuation Quantities from Time Series of Water Table Depths Using Models with a Stochastic Component. *Journal of Hydrology*, **197**, 25-46. http://dx.doi.org/10.1016/S0022-1694(96)03278-7

[21] Karamooz, M. and Aragi Nejad, S.H. (2010) Advanced Hydrology. 2nd Edition, Amirkabir University of Technology Press, Tehran, 464.

[22] Hashemi, R. and Jahanshahi, M. (2005) Analyze and Prediction of Annual and Monthly Total Precipitation in Torbat Hedarriye Region of Khorasan. *5th Seminar on Probability and Stochastic Process*, Birjand, 1-5 September 2005, 9.

9

Present Conditions and Future Challenges of Water Resources Problems in Iraq

Nadhir Al-Ansari, Ammar A. Ali, Sven Knutsson

Department of Civil, Environmental and Natural Resources Engineering, Lulea University of Technology, Lulea, Sweden
Email: nadhir.alansari@ltu.se, ammar.ali@ltu.se, sven.kuntsson@ltu.se

Abstract

Iraq is part of the Middle East and North Africa (MENA region). It greatly relies in its water resources on the Tigris and Euphrates Rivers. Iraq was considered rich in its water resources till 1970s. After that problems due to water scarcity aroused. Recently, it is expected that water shortage problems will be more serious. The supply and demand are predicted to be 43 and 66.8 Billion Cubic Meters (BCM) respectively in 2015, while in 2025 it will be 17.61 and 77BCM respectively. In addition, future prediction suggests that Tigris and Euphrates Rivers will be completely dry in 2040. To overcome this problem, prudent water management plan is to be adopted. It should include Strategic Water Management Vision, development of irrigation techniques, reduction of water losses, use of non-conventional water resources and research and development planning.

Keywords

Iraq, Tigris River, Euphrates River, Water Scarcity, Water Resources Management

1. Introduction

Iraq is part of the Middle East and North Africa (MENA region). It covers an area of 433,970 square kilometers populated by about 32 million inhabitants (**Figure 1**) [1] [2]. This region is characterized by its water shortage problem [1]-[6] where at least 12 countries have acute water scarcity problems with less than 500 m^3 of renewable water resources per capita available [7] [8]. The supply of fresh water is essential to life, socioeconomic development, and political stability in this region. It was reported that one cubic meter of water can provide drinking water for one person for one year and the same quantity can produce only one kg of food grain when used for irrigation in a dry climate [9]. The largest consumer of water across the region is agriculture where it accounts for 66% of demand [5] [10] and therefore the water shortage problem cannot be objectively analysed nor adequately

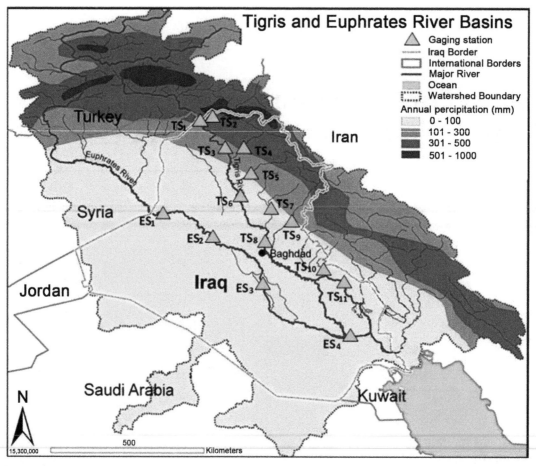

Figure 1. Physiography of Iraq.

addressed without a thorough consideration of agriculture [11]. If we consider 10% transfer of water away from agriculture would produce a 40% increase in domestic water supply for Jordan [11]. Postel [12] argues that rather than diverting precious water to agriculture this water could be saved by importing the food/grain. However, this is not the case in many Middle Eastern countries which have unrealistic aspirations of food self-sufficiency and it would require a most fundamental change in national outlook [13]. Sadik and Barghouti [11] emphasized that the extent of the problem is so severe that "the future challenges in meeting the growing demands for water are beyond the capabilities of individual countries". Mitigating supply shortfalls could be achieved by re-allocation of current agricultural supplies [3]. Future predictions suggest more shortages [14]-[16] and depletion of groundwater resources [2] [17] [18].

Iraq is bordered by Turkey from the north, 352 km, Iran from the east, 1458 km, Syria and Syria from the west, 605 km and 181 km, respectively and Saudi Arabia and Kuwait from the south, 814 km and 240 km respectively (**Figure 1**). Topographically, Iraq is divided into 4 regions (**Figure 1**). The mountain region occupies 5% of the total area of Iraq, restricted at the north and north eastern part of the country. This region is part of Taurus-Zagrus mountain range. Plateau and Hills Regions is the second region and it represents 15% of the total area of Iraq. This region is bordered by the mountainous region at the north and the Mesopotamian plain from the south. The Mesopotamian plain is the third region and it is restricted between the main two Rivers, Tigris and Euphrates. It occupies 20% of the total area of Iraq. This plain extends from north at Samara, on the Tigris, to Hit, on the Euphrates, toward the Gulf in the south. The remainder area of Iraq which forms 60% of the total area is referred to as the Jazera and Western Plateau.

Iraq was considered rich in its water resources compared with other countries where the annual allocation per capita reached 6029 m³ in 1995 and expected to be 2100 m³ in 2015 [19]. Construction of dams on the Tigris and Euphrates and their tributaries outside the border of Iraq, the effect of global climate change and mismanagement of water resources are the main factors in the water shortage problems in Iraq [1]. Restoring the marshes [20] [21] and the growing demand for water in Turkey and Syria will lead to dry the Tigris and Euphrates Rivers in 2040 [20]. Furthermore, the supply will be 43 and 17.61 BCM in 2015 and 2025 respectively while current demand is estimated between 66.8 to 77 BCM. These suggest that the Iraqi government need to take quick, prudent and firm action as a high priority. One of the solutions is the use of Water Harvesting techniques [1]. To overcome these problems in Iraq, there is a great need for prudent management of water resources and the adaptation of non-conventional techniques to augment water resources [1] [22].

In this research, the present conditions and future expectations are described to give ways and means to overcome the scarcity of water resources problems in Iraq.

2. Climate

The climate is mainly of continental, subtropical semi-arid type. The mountain region is of Mediterranean climate. In general, rainfall occurs from December to February or November to April in the mountain region. During winter the average daily temperature is about 16°C dropping at night to 2°C with possibility of frost. In summer however, it is very hot with an average temperature of over 45°C during July and August dropping to 25°C at night (**Figure 2**).

The annual rainfall in Iraq varies where it reaches 150 mm within the western desert, more than 1000 mm within the mountains at the north to about 200 mm at the eastern part of the country (**Figure 3**). The overall average annual rainfall is of the order of 213 mm per year. The rainy season begins in October and ends in April. It is evident from **Figure 3** that annual rainfall increase from southwest towards northeast due to topographic effect. Furthermore, it can be noticed that individual topographic regions are characterized by their own climatic factors and rainfall values. It should be mentioned however, that despite the local climatic differences, all the regions have similar overall climatic features. This is due to the fact that Iraq as a whole is affected by its geographic position. When rainfall is used as a base to classify the climate, two seasons can be noticed:

a) Dry season starting in April to September.

b) Wet season starting in October to May.

It can be stated that October represent the transition period from dry to wet seasons while May representing the transition from humid to dry season.

The range of daily temperature varies greatly between day and night reflecting the continental climate. In addition, the trend in temperature increase is exactly the opposite to that of the rainfall's trend, where it increases from northeast toward southwest. The maximum daily temperature during dry season could rise to over 50°C, while the minimum temperature during the wet season could reach −14°C in Rutba and about −8°C in Baghdad (**Figure 2**). It can also be noticed at the south, in Basrah in particular, that the temperature during summer is less than that of the surrounding areas due to high humidity resulting from being near to the Gulf.

Sunshine records indicate that during dry period, May-September, the average is more than 500 cal/cm²/day while it is below this value during wet season (October-April). Meteorological record was used to calculate the

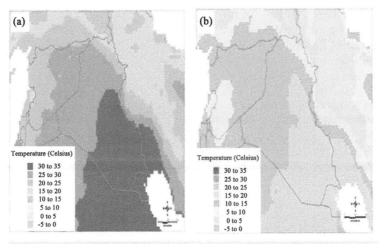

Figure 2. Mean maximum (a) and minimum (b) temperatures in Iraq.

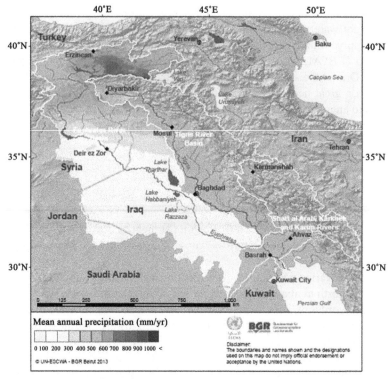

Figure 3. Rainfall map of Iraq [2].

evaporation and evapotranspiration values using penman method. The results show that the overall average evaporation and evapotranspiration is of the order of 1900 mm per year. Furthermore, the values show an increasing trend similar to that of the temperature increasing from northeast towards the southwest.

3. Water Resources

3.1. Surface Water

The rivers Tigris and Euphrates with their tributaries form the main surface resources in Iraq. Details of these basins can be found in Al-Ansari [1] (2013), Al-Ansari and Knutsson [22] and ESCWA [2]. The catchments area of these rivers is shared by five countries: Iraq, Turkey, Iran, Syria and Saudi Arabia, **Table 1** and **Figure 4**.

Several ancient civilizations in the Mesopotamia were supported by basin irrigation from the Tigris and Euphrates Rivers since 5000 BC during Sumerian time. The ancient irrigation system was so efficient where it support wide spread cultivation of the land for many years without serious decline in land quality. Due to these marvelous water activities the term "hydraulic civilization" was used to describe this society.

Generally, the total annual flow of the Tigris and Euphrates Rivers is of the order 80×10^9 m^3. This figure greatly fluctuates from year to year (**Figure 5**). Furthermore, floods and drought are themselves of variable magnitude. Such variations are due to changing metrological conditions. The period extending from October to February is referred to as variable flood period where discharges in both rivers fluctuate depending on intensity and duration of rainfall at their basins. This period is usually followed by what is known as steady flood period extending from March to April.

Flow records show that minimum and maximum annual flow of the Tigris was 19×10^9 m^3 in 1930 and $10^6 \times 10^9$ m^3 in 1969, while for the Euphrates it was 9×10^9 m^3 in 1974 and 63×10^9 m^3 in 1969, respectively. **Table 2** and **Table 3** summarize the source and uses of water in both rivers.

3.1.1. River Euphrates

River Euphrates rises in the mountains of southern Turkey. It runs about 1178 km in Turkey before interring the Syrian territory where it runs 604 km to reach the Iraqi border. Inside Iraq the length of river is 1160 km. the total length of the river from Turkey to its confluence with the River Tigris south of Iraq is 2940 km, the total drainage

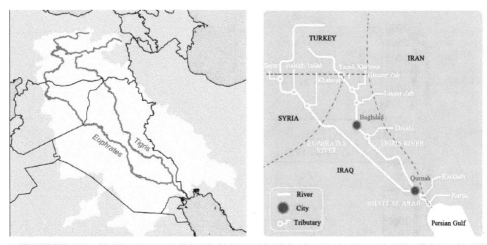

Figure 4. Catchments area of Rivers Tigris and Euphrates [2] [23].

Table 1. The area of Tigris and Euphrates Basins.

Countries	Tigris River		Euphrates River	
	Catchment Area (km²)	Catchment Area (%)	Catchment Area (km²)	Catchment Area (%)
Turkey	57,614	12.2	125,000	28.2
Syria	834	0.2	76,000	17.1
Iraq	253,000	58	177,000	39.9
Iran	140,180	29.6	-	-
Saudi Arabia	-	-	66,000	14.9
Total	473,103	100	444,000	100

Table 2. Sources and uses of the Euphrates River (million cubic meters (MCM) per year).

Natural Flow	Observed at Hit, Iraq	29,800
	Removed in Turkey (pre-GAP)	820
	Removed in Syria (pre-Tabqa)	2100
	Natural flow at Hit	32,720
Pre-Kaban Dam (before 1974)	Flow in Turkey	30,670
	Removed in Turkey	(820)
	Entering Syria	29,850
	Added in Syria	2050
	Removed in Syria	(2100)
	Entering Iraq	29,800
	Added in Iraq	0
	Iraqi Irrigation	(17,000)
	Iraqi return flow (est.)	4000
	To Shatt al-Arab	16,800
Full Use Scenario (circa 2040)	Flow in Turkey	30,670
	Removed in Turkey	(21,600)
	Entering Syria	9070
	Removed in Syria	(11,995)
	Return flow and Tributaries (Turkey, Syria)	9484
	Entering Iraq	6559
	Removed Iraq	(17,000)
	Return flow in Iraq	4000
	Deficit to Shatt Al-Arab	(6441)

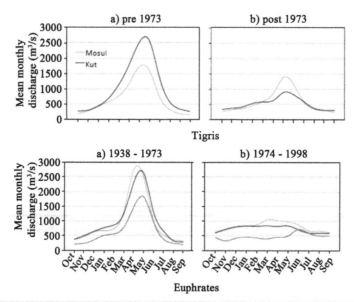

Figure 5. Mean monthly flows of the Tigris and Euphrates Rivers at different gauging stations for different time periods [2].

Table 3. Sources and uses of the Tigris River (MCM per year).

	Pre-GAP Project	Post GAP 2000 AD	Natural Flow
Flow From Turkey	18,500	18,500	18,500
Removed in Turkey	0	6700	
Entering Iraq	18,500	11,800	
Inflows to Mosul	2000	2000	2000
Greater Zab	12,100	13,100	13,100
Lesser Zab	7200	7200	7200
Other	2200	2200	2200
Sub-Total	43,000	36,300	43,000
Reservoir evaporation	0	(4000)	
Irrigation (to Fatha)	(4200)	(4200)	
Return Flow	1100	1100	
Adhaim	800	800	800
Irrigation(to Baghdad)	(14,000)	(14,000)	
RETURN Flow	3600	3600	
Domestic Use	(1200)	(1900)	
Diyala River	5400	5400	5400
Irrigation	(5100)	(5100)	
Return Flow	1300	1600	
Sub-Total	30,700	19,600	49,200
Reservoir evaporation	0	900	
Irrigation to Kut	(8600)	(8600)	
Return Flow	2200	2200	
		(to outfall drain)	
Total Shatt Al-Arab	24,300	14,100	49,200

area of the river catchments is 444,000 km², **Table 3**, of which 28%, 17%, 40% and 15% lie in Turkey, Syria, Iraq and Saudi Arabia, respectively.

The Euphrates River rises east the Anatolian plateau between Wan lake and Black Sea in Turkey with two main tributaries (**Figure 4**):

a) Furat Su: This forms the northern tributary where it rises from the mountains situated northeast Ardhroom area with altitudes varying from 1800 to 3937 m above sea level. The total length of the tributary reaches 510 km.

b) Murad Su: This forms the southern tributary where it rises north of Wan Lake at areas having altitude of 2350 to 3519 m above sea level with a total length of 600 km.

Both tributaries join together 5 km north of Keban city. At Keban area the Keban dam was constructed with a total reservoir capacity of 30.5×10^9 m^3. The average annual flow of the Euphrates River in this area is 672 m^3/s. The united river length in Turkey is 455 km. The shape of the catchments of the river in Turkey is a fan shaped which collects runoff at rainy periods in a short period of time causing sudden flood peak. The Euphrates River inters the Syrian border near Jorablus and flow to Albokamal at the Syrian-Iraqi border. The catchment's area of the river before entering Iraq reaches 201,000 km^2.

In Syria, three tributaries join the Euphrates:

a) Sabor River: total length of the river is 108 km, joining the Euphrates, from the right, 30km south of Tripoli city. The average discharge of the river is 3 m^3/s.

b) Belaikh River: The length of this river is 105 km and joins the Euphrates from the left south of Raka city downstream Tabaka dam. The average discharge of the river is 36 m^3/s and its catchment area is 14,400 km^2.

c) Khabor River: The length of this river is 446 km. Its catchment area is 36,900 km^2 and lies in Turkey and Syria. Four small streams, *i.e.* Jaja, Jabjab, Etehad Aracha Alsaghir and Etehad Aradha Alkabeer, join this river. The khabor join the Euphrates south of Dier Al-Zor city and its mean daily discharge is 55.8 m^3/s and could reach 500 m^3/s during flood.

There are several dams constructed on the Euphrates River in Syria. One of the main dams is called "Tabaka". The storage capacity of this dam is 11.6×10^9 m^3 (life storage 7.4×10^9 m^3). The maximum, minimum and mean discharge of the river Euphrates at Tabaka dam is 8500, 450 and 1300 m^3/s, respectively. There are three more large dams; Tersanah, Teshreen and Muhardah with storage capacities of 225×10^6 m^3, 210×10^6 m^3 and 50×10^6 m^3, respectively. In addition, 84 other small and medium size dams exist. The largest among these dams is known as Babalhadied with a storage capacity of 25×10^6 m^3, while the storage capacity of the smallest dam is 30,000 m^3.

The length of the river Euphrates in Iraq from the Syrian-Iraqi border at Hussaybah to its confluence with the Tigris River is 1160 km. once the river enters Iraq it trends toward the east and southeast to reach Anah city 100 km south of Hussaybah. The river then runs 220 km to reach Hit, with a river channel slope of 1:320 m. The channel is characterized by its shallow depth, large width and meanders. Islands are also noticed at the river channel. No tributaries join the Euphrates River inside Iraq apart from dry valleys originating from western desert. These valleys supply the river with flood water at rainy season from the desert. The minimum and maximum recorded discharges at Hit are 55 on 5/9/1973 and 7460 m^3/s on 13/5/1969.

The river runs 63 km south Hit to reach Ramadi city with a width reaching 250 m. At this area, Ramadi barrage was constructed to supply Warrar stream with water. This stream supplies Habariya Lake with excess water from the Euphrates during flood period. The length of this stream is 8.5 km which is designed to discharge up to 2800 m^3/s. Water stored in Habaniya Lake can brought back to the river Euphrates when required through Dhiban stream which is located 42 km south of Ramadi city. In case of continuous flood flow, Habaniya lake cannot accommodate huge volumes of water and thus it was connected to Razazah lake and then to Abudibis marsh to release excess water.

South of Ramadi city the river flows 72 km to reach Falujah city. A complex of canals system was constructed at this section of the river. A canal was dug from Tharthar large lake to supply the Euphrates River during drought periods; the canal joins the river 35 km north of Falujah city. The canal is designed to discharge water up to 1100 m^3/s. It has a diversion which can supply excess water to the Tigris River if required. This diversion canal is 65 km long.

The Euphrates River runs south about 110 km to reach Hindiyah barrage. The river runs through fluvial deposits in this area. It should be mentioned however, that the Euphrates River channel inside Iraq upstream Hindiyah is higher by 7 m than Tigris river channel. This is due to the fact that the Euphrates runs on the edge of the western plateau. This phenomenon was used since ancient times to dug irrigation canals from the Euphrates River running toward the Tigris River.

At Hidiyah, the river is known as Hindiyah River where it flows south to reach Kifil city 18 km south of Hindiyah barrage. South of Kifil, the river splits to eastern channel (Shamiyah) which takes 40% of the flow and western channel (Kufa) taking the remaindering 60% of the flow. On shamiyah river there are various regulating schemes established for a number of irrigation projects, *i.e.* Danaieb Alshanafiyah, Shalal Danaieb Alshanafiyah, Khuman and Naghshiyah regulating schemes. This channel joins the Euphrates River again 8 km upstream Shanafiyah city. The distance from Kifil to Shanafiyah is 99 km.

As far as the other channel, the Kufa, it runs to Abusskhair city and splits into two main channels. The small channel (on the right of the main channel) supplies four small irrigation streams, the main channel south of Abusskhair as Mushkhab River where Mushkhab barrage was established. The water flows to the south to reach Qadisiyah city and it again splits and joins again to form one channel. This channel joins Shamiyah channel again at Shanafiya city.

The river continues to run south for 105 km where it reaches Simawah city. After that the river Euphrates reaches Nasiriyah city. It should be mentioned however that this section of the river (Simawah-Nasiriyah section) has a number of irrigation projects and small intake stream. South of Nasiriyah city the river runs toward Al-shiyokh city and then enters Hammar Lake. Two channels leave Hammar Lake. The first joins the Tigris River at Qurna City while the other, the southern one, joins Shatt Al-Arab River at Karmat Ali.

3.1.2. River Tigris

The total drainage area of the Tigris River is 235,000 km^2 distributed between Turkey (17%), Syria (2%), Iran (29%) and Iraq (52%) (**Figure 4**). The overall length of the river is 1718 km. It rises at the southeastern slope of Taurus Mountains at two sites, the western site is located near Diar city, 1500 m above sea level, with a discharge of 64 m^3/s while the eastern site, known as Butman Su, near Sinan city a 2700 m above sea level with a discharge of 96.3 m^3/s. The river runs in a narrow valley bounded by Mardin Mountain range from the right and Raman Surat Hills from the left. Further down, another tributary (Karzan) joins the river near Bishwi village. The river runs south through rough mountainous area till it is joined by another tributary (Hazu) near ZEU village 240 km north of the Iraqi border. The discharge of this tributary is 59 m^3/s. The river then runs in a plain area and joined by Butan Su River, with a discharge of 20.3 m^3/s, to form the united Tigris River.

The river enters Iraq 4 km north Fieshkhabur near Zakha city. The Tigris is joined by its first tributary inside Iraq which is known as Khabur River. This tributary is 100km long, it catchment area is 6268 km^2 with an average, maximum and minimum discharge of 68 m^3/s, 1270 m^3/s (11/4/1963) and 8 m^3/s (6 - 14/9/1962), respectively. The Tigris River runs south for about 188 km in a hilly area to reach Mosul city. At Mosul the average, maximum and minimum discharge of the river is 668 m^3/s, 7740 m^3/s (2/5/1972) and 85 m^3/s (October, 1935) respectively. The elevation of the channel bed is 225 m above sea level.

About 49 km south of Mosul toward Sharkat city, the Tigris joins its biggest tributary, the Greater Zab. The catchment of this tributary lies in Turkey, Iran and Iraq. Its total length is 437 km with a mean, maximum and minimum discharges of 450 m^3/s, 9710 m^3/s (2/4/1969) and 60 m^3/s (22/11 and 4/12/1958) respectively. It supplies 28.7% of the Tigris water.

The Tigris River runs south toward Fatha gorge. About 30 km north of Fatha, the Lesser Zab tributary joins the river Tigris. The total catchment area of this tributary is 22,250 km^2 of which 5975 km^2 lie in Iran and the remainder in Iraq. The total length of the river is 456 km. The mean, maximum and minimum discharges of Lesser Zab are 227 m^3/s, 3420 m^3/s on 8/3/1954 and 6 m^3/s on 14/5/1964, respectively. The river Tigris mean, maximum and minimum discharges at Fatha gorge are 1349 m^3/s, 16,380 m^3/s on 3/4/1969 and 200 m^3/s in October 1930, respectively.

Further south, the Tigris River enters the Mesopotamia plain 20 km north of Samara city and then Balad city. The Adhaim tributary joins the Tigris 15 km south Balad. The tributary drains an area of 13,000 km^2 lying within Iraq. Its length is 330 km. The mean daily discharge is of the order of 25.5 m^3/s; while the maximum recorded on 19/10/1960 reached 3520 m^3/s. this tributary runs almost dry during June to November each year. The banks of the river Tigris south its confluence with Adhaim tributary are below the maximum flood peak level by 3 m from the left and 1.8m from the right.

Further to the south, the river reaches Baghdad. At Baghdad the mean, maximum and minimum discharges are 1140 m^3/s, 7640 m^3/s on 12/2/1941 and 163 m^3/s in October, 1955 respectively. The slope of the channel is very low *i.e.* 6.9 cm/km. About 31km south of Baghdad, the last main tributary "Diyala" joins the Tigris. Diyala's drainage basin is 31,896 km^2 of which 20% lies in Iran and the rest in Iraq. The mean daily discharge is 182 m^3/s while the maximum and minimum discharges are 3340 m^3/s on 25/3/1954 and 12 m^3/s on 7/9/1960, respectively.

Downstream the confluence of the Tigris-Diyala River, the Tigris channel is characterized by its large number of meanders. In addition, the river discharge steadily decreases downstream due to losses. These losses include evaporation, infiltration and mainly water withdrawal through irrigation canals. Important irrigation canals includes: Gharaf, south Kut city, Great Majar, South Emarah city, Mushrah and Kahlaa, South Emarah city, and

Majariah canal, South Qalaat Salih. In addition, this stretch of the river is well known by the abundant occurrence of big marshes on both sides of the Tigris River. There are many small streams running from Iran toward Iraq where they discharge their water in the marshes e.g. Karkha stream discharging its water in Hiwazah marsh.

The Tigris channel reaches its minimum width at Kasarah area south of Emarah city. At Qalaat Salih the mean daily discharge of the river is 80 m^3/s. Downstream this city the river joins the Euphrates River at Qurnah city forming Shatt Al-Arab River.

3.1.3. Shatt Al-Arab River

Shatt Al-Arab River is formed after the confluence of Tigris and Euphrates Rivers at Qurnah in Iraq (**Figure 6**). Its total length is 192 km and its drainage area is 80,800 km^2. Its width is about 300 m near Qurnah and increases downstream to 700 m near Basra city and to about 850 m near its mouth at the gulf area. Karun and Karkha Rivers usually contributes 24. 5 and 5.8 billion cubic meters (BCM) annually respectively (**Figure 6**). This forms about 41% of the water of Shatt Al-Arab. Its annual discharge at Fao city reaches 35.2×10^9 m^3. Shatt Al-Arab River is characterized by its high sediments which resulted in the formation of large number of islands during its course.

3.2. Groundwater

Kahariez mountain channels and hand dug wells were common practices in the ancient history of Iraq. Some of these wells and Kahariez are still in use now. Large numbers of villages were previously built near springs and later wells were dug to provide water for the inhabitants. The geographic distribution of springs and wells marked the travel routes in ancient Iraq. Development and utilization of groundwater started in 1935 where the first groundwater well was mechanically drilled. This development went through four main stages:

I. Stage one: this stage started in the mid-1930s to the beginning of the 1950s. During this stage wells were drilled without any scientific investigation or studies. Drilling operations were executed to provide water for villages and remote areas. Few basic reports on groundwater resources were written during this stage.

II. Stage two: this stage started during the 1950s toward the end of the 1960s. This stage is marked by the activities of foreign consultants and contractors. Companies like Parsons and INGRA carried out countrywide surveys for groundwater resources in Iraq. A numbers of wells were drilled to for the purpose of the study required. This operation resulted in a huge survey report on the groundwater resources of Iraq published in 13 volumes.

III. Stage three: this stage started at the end of the 1960s. The most important feature of this stage was the availability of Iraqi geologists and drilling engineers to drill the required wells all over Iraq. This stage was relatively poor in its scientific research and studies where field drilling operations were predominant and hundreds of wells were established.

IV. Stage four: this stage started during the 1970s and marks scientific research for groundwater investigations and utilization. This was executed by national and foreign organizations and companies. The Ministry of Irrigation (now Ministry of Water Resources) was the official body responsible for these operations and practices. During this period the state company for water wells drilling was established.

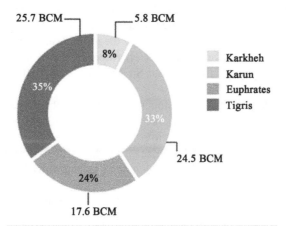

Figure 6. Water contribution to Shatt Al-Arab River [2].

Despite the large number of groundwater wells that exists now, ground water utilization in Iraq forms a minor percentage (5% - 7%) of the water resources of the country despite its extensive use [24]-[32]. Ground water resources in Iraq are described in details by Al-Ansari [24], Al-Sayab *et al.* [25], Al-Ansari *et al.* [26]-[30], Alsam *et al.* [31] and Kransy *et al.* [32]. However, using the geological, structural and physiographic characteristics, groundwater in Iraq can be summarized by divided to five main zones:

3.2.1. Mountain Region

This region is confined to a small area, 20,000 km^2, in the north and northeastern part of Iraq. It is characterized by the availability of groundwater resources. This region can be divided into three sub regions on the basis of petrology of the aquifers, their age and distribution.

1) Sub region one: this sub region covers limestone, dolomites and other hard rocks. The age of the rocks is cretaceous or older. The rocks of this area are characterized by their joints and fissures where groundwater is usually found within these features. The quality of groundwater is good and can be used for different purposes. The reserve is considered medium. The groundwater in this region is usually utilized by springs.

2) Sub region two: This sub region is characterized by its weak sedimentary rocks e.g. sandstones, shale and marl, etc. The age of the rocks is cretaceous where fine grained sandstones are the principle aquifers in the area as well as marls and siltstones, where the rocks are usually jointed. The aquifers are usually of the confined type and distributed within the valleys at different depths, there is no detailed exact study regarding the quality of groundwater but it had been estimated to be medium and of good quality. Water is usually utilized through springs, hand and mechanical dug wells.

3) Sub region three: The rocks in this sub region are composed of Pleistocene and Quaternary recent valley deposits. Gravel, conglomerates, sand and sandstones form the main aquifers. Groundwater is found at shallow depths (about 30 m) with very good quality (salinity 170 - 350 ppm) and with adequate quantities. Hand dug wells, mechanical wells, springs and kahariez are the methods of groundwater utilization in this area. Kalatdiza, Rania and Shahrzor are the main important groundwater basins within this sub region.

3.2.2. Plateau and Hill Region

This region border's the mountain region and covers an area of 62,000 km^2. The main aquifers are of Pliocene and Pleistocene age (sometimes Miocene). The aquifers of this region are considered to be of the highest quality and have the best quantity. The hills are usually folded strata and are parallel extending from NW-SE. the elevation of these hills is about 200 m in the south while it increase's to 500 m in the north. These hills are usually narrow with very wide plains between adjacent regions. This region can be sub divided in to two sub regions, according to the rock type of the aquifers their age and distribution:

A. Sub region one: The main aquifers are Bakhteiari Formation (Pliocene-Pleistocene) and recent alluvial deposits. The aquifers are composed of conglomerates, sandstones, sand and gravel. The thickness of the aquifer varies from place to place and sometimes reaches 400 m. Groundwater depth does not exceed 50 m below ground level and saturated thickness reaches 400 m. The salinity of groundwater varies from 300 ppm (Irbil basin) to 2500 ppm (Mandeli and Wend). The main basins are:

- Irbil plain: This plain extends between the L-Zab River at south and the G.Zab River at north with a total area of 2000 km^2. The aquifers are composed of conglomerates, sandstone, sand and gravel. The salinity range is 300 - 500 ppm. The yield of the wells is 10 - 30 l/s while the groundwater level is 25 - 40 m above sea level.

- Altonkupri plain: This plain is 20 km north of KirkEU, bordered by the Klilkhal Mountain at the northeast, the L-Zab River from the north-northwest and the Khasa Su River from the south east, the quality of water is good and the yield of the wells varies from 7 to 30 l/s.

- Aqra plain: This plain is bordered by the Aqra Mountain from north, the Khabour River from the west, the G. Zab River from east and the Maklob Mountain from the south. The quality of water is very good (salinity range 300 - 500 ppm) and the yield of the wells varies from 10 - 30 l/s.

- Wen River basin: This basin extends from Khanaqien city at the northeast toward the confluence of the Wend River with the Diyala River at the southwest. The quality of groundwater is good to medium (salinity 500 - 2000 ppm) while the yield of wells is 5 - 20 l/s.

B. Sub region two: Groundwater in this area is found in Fars Formation beds. The quality and quantity of groundwater is considered medium to poor. The aquifers are mainly sandstones and shales. The most important basins are:

- North Sinjar plain: these basin extents from Sinjar Mountain at the south towards the Iraqi-Syrian border at the north covering an area of 1360 km². The salinity of water is 300 - 2500 ppm but it ranges 3000 - 13,000 ppm in the southern parts of the basin. The yield of the wells is of the order of 7 l/s.
- Rabia plain: Groundwater aquifers are restricted to sandstone bed of the Upper Fars Formation. The salinity range is 500 - 1000 ppm, while it ranges 1000 - 5000 ppm in the southern and eastern parts of the basin. Wells yield range 7 - 12 l/s.
- Tikrit-Sammara basin: This area is bordered by Hemrin and Makhul mountain ranges from the northeast, the Tharthar depression from the west and the Adhaim River from the east. The salinity range 2000 - 5000 ppm whiles the yield of the wells range 7 - 10 l/s.

3.2.3. Mesopotamian Plain (Delta) Region

This region covers the area bounded by the Rivers Tigris and Euphrates south of Baghdad to the Arabian Gulf. The groundwater in this region is usually found within the recent alluvial deposits. The Tigris and Euphrates are the main source for groundwater in this region. The salinity varies with depth where it is of the order of 3500 ppm at depths less than 20 m while it reaches 20,000 ppm at greater depths, the groundwater table is very shallow near the ground level. In restricted areas fresh water of very good quality can be found overlying saline groundwater.

3.2.4. Jezira Plain Region

This region occupies the northwestern part of Iraq, bordered by Sinjar mountain from the north, the Tharthar depression from the east, the Euphrates river from the south and the Iraqi-Syrian border from the west the area is characterized by its almost horizontal beds, where the Upper and Lower Fars Formations are usually exposed on the surface. These formations usually act as aquifers. The water quality of this region is usually poor due to high salt content from Lower Fars beds. Groundwater quality can be considered medium (salinity 3000 - 5000 ppm) within the Upper Fars beds west of the Sinjar area but salinity increases toward the south.

3.2.5. Western Desert region

This region covers an area of about 226,000 km². It is bordered from the northwest by the Iraqi-Syrian border, from the west by the Iraqi-Jordanian border, from the southwest by Saudi Arabia, from the southeast by Kuwait and the Euphrates River from its northern border. This region forms a plain area with no marked physiographic features. The slope of the area is directed from the Iraqi-Saudi Arabia border in the south west towards the Euphrates River in the north east; and from the Iraqi-Syriaian border in the west toward the Euphrates River in the east. The maximum altitude reaches 900 m above sea level near the Iraq-Syria border while the minimum is less than 50m above sea level near the Hamar marsh.

The area is characterized by its large number of seasonal streams and lakes. The Khir valley divides the region into what is known as the north Badia and the south Badia. Groundwater forms the main source of water in this area and it can be found in different geologic formations. Generally groundwater movement is toward the northeastern part of the desert where in some places it runs to the surface in the form of springs.

Using the type of aquifers and its age the area can be divided into three main sub regions:

Sub region one (Dibdibah):

This area is confined to the extreme south eastern part of the region. The main aquifer is the Dibdibah Formation (Pliocene-Miocene) which is mainly composed of sand, gravel and gypsum. The thickness of this formation varies from place to place and it can reach 140 m.

Groundwater salinity is relatively high (2500 - 8000 ppm). Generally shallow groundwater (top 10 m of the saturated thickness) is of a better quality relative to deeper water. Yield of the wells ranges from 5 to 10 l/s. Utilization of groundwater is extensive in this area for agricultural purposes especially near the Zubair-Safwan area. Despite the high salinity of the water, agricultural activities are very successful which might be due to the light and higher porosity of the soil.

Sub region two (Calcareous rock area):

This area covers most of the western desert region and is characterized by its calcareous rock of different geologic ages (Miocene to Tertiary). The most important formations are: Dammam, Umerdhumah, Tagarat, Masad and Euphrates.

The rocks of the area are jointed and fractured with thicknesses reaching 400m. Groundwater occurs at a depth of 300 m and the aquifers are usually unconfined except the eastern parts where they are of a confined type. The salinity range is 2000 - 5000 ppm. The yield of the wells is usually 4 - 25 l/s. the most promising basins for groundwater utilization are:

- Salman basin: this basin covers 35,000 km^2 in the southern desert. Three main aquifers are found in this basin (Dammam, Umerdhumah and Tayarat). Groundwater depth is usually 50 - 130 m while the salinity is 2000 - 8000 ppm. The yield of the wells is 3 - 10 l/s.
- Hawar-Muainah basin: This basin is located within the southern desert. The aquifers are composed of limestone and dolomite. Groundwater depth is 75 - 110 m with salinity range 2800 - 3500 ppm. The average yield of well reaches 15 l/s.
- Arar Valley basin: The area of this basin is 2800 km^2. Umerdhumah and Tayarat Limestone beds form the main aquifers of this basin. The salinity is 700 - 3000 ppm and the yield of wells is 25 - 30 l/s.
- WadiKhir basin: The area of this basin is 3000 km^2. The main aquifers are Dammam and Umerdhumah Limestone beds. The salinity is 1700 - 3000 ppm and wells yield is 30 l/s.
- WadiHamer basin: This basin covers 3400 km^2 in the northern desert. The main aquifer is Tayarat Limestone Formation. The salinity is 500 - 3000 ppm and well yield varies from 20 to 25 l/s.
- Kasra area basin: This basin is located in the northern desert about 125 km west of Ramadi city. It extends between the Ghadof valley at the north and the Abyiadh valley at the south. The basin is 75 km in length and 50 km in width. Tayarat Limestone Formation is the main aquifer. The quality of water is excellent to very good, salinity 160 - 1000 ppm and the average yield of wells is 25 l/s.

Sub region three (Sandstone area):

This sub region is characterized by cretaceous and older rocks and covers a wide area within the northern desert. Groundwater is found in sandstone and limestone beds at depths of 50 - 300 m. The yield of wells varies from less than 1 l/s to 20 l/s in confined and unconfined aquifers. The main aquifers are: Mehawir, Ubaid, Mulusa, Qaara and Sufi.

Qaara Formation (Upper Permian - middle and lower Triassic) is the most important aquifer with a varying thickness from 180 to 720 m. This formation contains two aquifers. The first is 150 - 450 m deep. The yield of this aquifer is 2 - 10 l/s with salinity ranging 550 - 3000 ppm. The second aquifer is 700 m deep with high groundwater yield, up to 20 l/s. The salinity is of the order of 3000 ppm. It should be mentioned however that the most promising areas for water utilization is the Qaara depression and Turaibiel.

Despite the presence of the great Tigris and Euphrates Rivers, groundwater is considered the only source of water in the western desert and 70% of water consumed by villages in north Iraq. Thousands of wells, **Table 4,**

Table 4. Number of groundwater wells drilled in each governorate in Iraq.

Governorate	Number of Wells
Duhok	410
Naunawa	1299
Erbil	1286
Sulaymaniah	423
Tameem	1093
Diyala	647
Salahaldin	1118
Baghdad	308
Anbar	608
Muthana	201
Qadisiya	6
Karbala	148
Najaf	286
Wasit	116
Mesan	80
Dhiqar	17
Basra	576
Babil	30
Total	**8752**

were drilled in different parts of Iraq for various purposes, especially in the 1980s where a large number of wells were drilled for agricultural purposes.

The total number of wells drilled up to 1990 reached 8752 of which 1200 are used for agricultural purposes. These wells were drilled by the government. The number of wells drilled by the private sector reached 400. After the Iraq-Kuwait war no records are available about the number of wells drilled. It is believed that a large number of wells drilled during the 1990s where the government encouraged the private sector to increase agricultural productivity due to UN sanctions.

It is believed that groundwater utilization will be tremendously important in Iraq in the near future due to:

About 90% of the irrigation practice in Iraq depends largely on Tigris and Euphrates Rivers. These activities are confined to areas within the vicinity of these rivers and the Mesopotamian plain. This leaves about 60% of the total area of Iraq where surface water is not available. This fact will inevitably increase the importance of groundwater utilization to secure food for the continuous increasing population in the country.

The increasing water projects in Turkey, e.g. GAP, and Syria's intention of reducing the flow of the Tigris and Euphrates to Iraq in the near future. One of the solutions to overcome this crisis is to increase groundwater utilization.

There are increasing problems with regard to the allocation of water from surface irrigation projects between the farmers and the time water is released to the different farms. These problems led some farmers to drill their private wells to maintain the required water at the required time.

The estimated total water use is about 12 billion cubic meters. More than 50% of the ground water is consumed for domestic purposes (**Table 5** and **Table 6**) [31].

Alsam *et al.* [31] also calculated exploitable groundwater resources. The quantity of ground water with salinities of 1 - 3 and 5 - 10 g/l that can be used for irrigation and watering livestock about 3.8×10^9 and 2.25×10^9 m^3/year respectively. The quantity of more saline ground water (10 g/l) that can be used for industrial purposes is about 1.54×10^9 m^3/year (**Figure 7**).

In general, investment and optimal use of groundwater in Iraq is still at the beginning stage and does not exceed 5%.

Table 5. Water resources of main aquifer zones of Iraq [31].

Aquifer No.	Total Area Km^2	Modulus l/s km^2	Resource l/s	Less than 1 g/s		1 - 3 g/s		3 - 5 g/s		5 - 10 g/s	
				Area km^2	Resource l/s	Area km^2	Resource l/s	Area km^2	Resource l/s	Area km^2	Resource l/s
1	12,300	0.75	9225	0	0	0	0	0	0	0	0
2	1000	0.5	5000	3300	1650	0	0	0	0	0	0
3	43,300	0.5	21,650	6400	3200	0	0	0	0	0	0
4	31,800	0.5	15,900	0	0	6300	3150	5600	2800	0	0
5	61,200	0.35	21,420	0	0	31,700	11,095	900	315	0	0
6	18,400	0.25	4600	7700	1925	0	0	0	0	0	0
7	21,900	0.4	8760	1300	520	13,100	5240	5600	2240	1000	400
8	6400	0.5	3200	0	0	2700	1350	1500	750	2200	1100
9	40,100	1	40,100	0	0	21,700	21,700	11,300	11,300	2900	2900
10	79,300	0.75	59,475	0	0	200	150	23,400	17,550	55,500	41,625
11	37,600	2	75,200	0	0	10,800	21,600	14,600	29,200	1700	3400
12	10,600	3	31,800	3100	9300	1200	3600	1600	4800	0	0
13	40,700	2.5	101,750	24,300	60,750	2000	5000	1200	3000	0	0
14	24,500	5	122,500	24,500	122,500	0	0	0	0	0	0

Table 6. The total dynamic reserves of water in three categories (domestic, irrigation and livestock and unusable) [31].

	Total Amount of Ground Water (Billions of Cubic meters) and Water Use			
Sub Province	Aquifer No	Domestic1 g/s	Irrigation & Livestock 1 - 3 g/s	Unusable 3 - 5 g/l
Western desert	1, 2, 3, 4, 5, 7, 8	0.23	1.66	0.63
Jezera	9	0	0.13	0.68
Baiji-Tib	11	0	0.55	0.92
Baghdad Basra	10	0	0	0.004
Foothill zone	12, 13	2.21	1.48	0.270
High folded zone	14	3.86	0	0
Total	-	6.3	3.82	2.504

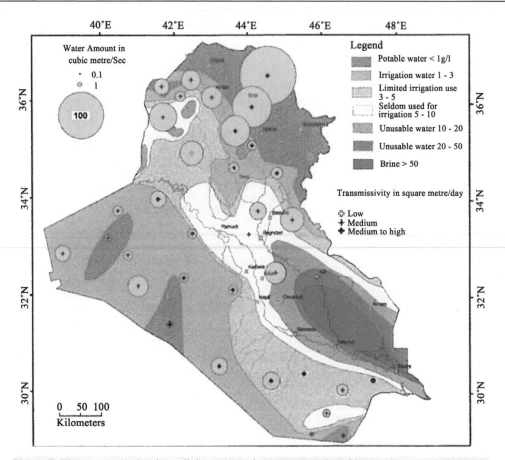

Figure 7. Water use map showing salinity zones and water amount in cubic meters per second as scale points [32].

3.3. Irrigation Projects

Irrigation, since the dawn of civilization about 7000 years ago has been an important base for agriculture in Mesopotamia, known as Iraq now. The only source for irrigation water in Iraq was from the Tigris and Euphrates rivers. The nature of these rivers' topography and climate imposed great problems on irrigation practices since its inception. These rivers are renowned for their dramatic spring floods and tremendous amounts of silt they carry. Furthermore, the plains of the country are very flat and poorly drained. As a consequence, the region suffered from persistent problems of poor soil, drought, and catastrophic flooding, silting and soil salinity.

The flooding from the two rivers usually peaks in March and May. This is too late for winter crops and too early for summer crops. In addition, the flow of the rivers varies considerably every year. The floods are usually too intensive which instigated some scholars to base numerous flood legends on it, e.g. the Gilgamesh Epic. Conversely, years of flooding make irrigation and agriculture difficult. This made people to opt to water storage and flood control as well as irrigation. In order to overcome the difficulty of frequent flooding, irrigation canals were dredged regularly to overcome rapid silt build up threatening to choke these canals. A more insidious problem to overcome was the draining off of flood water which had the tendency for washing salt build up in the soil. To overcome this situation, the Euphrates River was used as a supply and the Tigris channel as drain.

During Sumerian time each city used to build its own irrigation canals. The cities were originally administrative centers, marketing centers and defensive centers all connected to local irrigation schemes. Through the past history of Iraq engineers built very large weirs and diversion dams to create reservoirs and to supply canals that carried water for long considerable distances.

The prosperity of irrigation projects continued through the Islamic period. During the period 762 AD to 1258 AD, irrigation schemes were renovated and greatly extended. Water was carried out of the River Euphrates and led into parallel canals across the Mesopotamia plain where huge areas were irrigated. There was extensive salinity in the south. When the central government began to fail about the twelfth century the canals became silt choked, irrigation systems deteriorated and the land became more saline. In 1200 AD massive flood destroyed the whole system where the Tigris and the Euphrates changed their courses, cutting off most of the water supply to irrigation canals. What remained of the system collapsed in 1258 AD when the Mongols devastated Iraq. Iraq has remained as a desert for more than 600 years.

Not until the twentieth century did Iraq make a concerted effort to restore its irrigation and drainage network and to control seasonal flooding. Different governments constructed several large dams and river control projects, rehabilitated old canals and built new irrigation schemes. Examples of major projects are given in **Table 7**. Barrages were constructed on both rivers to divert water into natural depressions so that floods could be controlled, In addition, water can be used for irrigation after the rivers peaked in the spring.

Table 7. Important water irrigation projects during the twentieth century.

	Project	Location	Year Established
1	Nadiem Basha Embankment	E. Baghdad	1911
2	Hindiyah Barrage	Hindiyah	1913
3	Submerged Diyala Weir	N. Mukdadiyah	1928
4	Fist discharge recording station	Baghdad	1930
5	Diayala Irrigation Project	N. Baghdad	1937
6	Kut Barrage	Kut	1939
7	Dokan Dam	L. Zab River	1954
8	Ramadi Barrage	Ramadi	1956
9	Musaiyab Irrigation Project	Musaiyab	1957
10	Derbendikhan Dam	Diyala River	1961
11	Damlage Irrigation Project	Damlage	1963
12	Dibis Dam	L. Zab River	1965
13	Ishaki Irrigation Project	Ishaki	1966
14	Small desert dams	W. Desert	1970
15	Saddam Irrigation Project	Kirkuk Gov.	1971
16	Great Drainage Canal	Central Iraq	1972
17	Husainiyah Irrigation Project	Husainiyah	1974
18	Kifil-Shanafigah Irrigation Project	Kifil-Shanafigah	1975
19	Shabja and Sawari dams	W. Desert	1976
20	Hemrin Dam	Diyala River	1985
21	Mosul Dam (Saddam Dam)	Mosul	1986
22	Qadisiyah Dam	R. Euphrates	1986
23	Duhok Dam	Duhok	1985
24	Adaim Dam	Adaim	1999

Some dams were built on the tributaries of the River Tigris and on the main rivers as well. The total storage capacity of those dams was 13.7 cubic kilometers in the 1970s. During the 1980s a dam construction program was adopted. Mosul dam (Saddam dam) on the Tigris was constructed with a total storage capacity of 11.1 cubic kilometers. Qadisiaya dam was built on the Euphrates. Desert dams were also built having a storage capacity of 0.5 cubic kilometers, several other dams, such as the Bakhma dam on the Greater Zab river with 17.1 cubic kilometers storage capacity and the Badash dam on the Tigris river with 0.5 cubic kilometers, were under construction till 1990. The construction stopped after the first Gulf war. In addition to these dams two storage lakes had been created. The Tharthar (storage capacity 85 cubic kilometers) and Habbaniya which can the filled and drains into the lower reaches of the Euphrates.

To increase water transport efficiency, minimize losses and water logging and improve water quality, a number of new canals were constructed mainly in the southern part of Iraq. The most important was the "Third River" which functions as a drainage water of more than 1.5 million hectares of agricultural land extending north Bashdo to the gulf . This river which runs between the Tigris and the Euphrates was completed in 1992 with a total length of 565 km and a discharge of 210 m^3/s.

Long term data show that Iraq allocates 92% of its water resources for the agricultural sector. The overall cultivated area in Iraq mounts to 5,450,000 hectares of which 2,300,000 hectares are irrigated land. During 1990, water withdrawal was estimated to be 42.8 cubic kilometers of which 92% was used for agriculture, 3% for domestic supplies and 5% for industrial use. Irrigation potential in 1990 was estimated at over 5.5 million hectares of which 63% was in the Tigris basin, 35% in the Euphrates basin and 2% in Shatt Al-Arab basin, As far as soil quality in concerned, 6 million hectares were classified as excellent, good or moderately suitable for flood irrigation. With the increase of the volume of regulated flow the irrigation potential increased significantly. The total water managed area reached 3.5 million hectares which is equipped with partial control irrigation. The areas irrigated from surface water are about 3.3 million hectares of which not more than 30% in Shatt Al-Arab, the remainder are within Tigris and Euphrates basins. It is noteworthy to mention that all these areas are not currently irrigated since a large part has been abandoned due to water logging and salinity. It had been estimated that only 1,936,000 hectares were actually irrigated in 1993.

As far as irrigation using ground water resources, it was estimated that 220,000 hectares were irrigated using 18,000 wells. Furthermore about 8000 hectares were equipped for micro-irrigation, but these techniques were not used. During the year 2000, the total renewable water resources in Iraq were 75.42 cubic kilometers. Irrigation water requirements were 11.2 cubic kilometers while water use efficiency in percentages was 28%. Water withdrawal for agriculture reached 39.38 cubic kilometers and water withdrawal as percentage of renewable water resources were 52%.

Strategic crop growing activities during the 1990s show that 224,490 hectares were of irrigated wheat with an average yield of 2.7 tons/ha, while rain fed wheat areas were of the order of 508,620 hectares, with an average yield of 1.7 tons/ha. As far as barely is concerned, there were 200,700 hectares of irrigated barely with an average yield of 1.8 tons/ha. Rain fed barely areas were 323,730 hectares with an average yield of 1.3 tons/ha. Other craps like rice, vegetables, and corn and date trees were also irrigated.

3.3.1. Irrigation Projects on the River Tigris
The most important projects are:
A. Existing projects:
- Hemrin Dam: This dam was constructed on the river Diyala 10 km upstream the Siddor weir. The storage capacity of the dam is 3.95×10^9 m^3 with a total drainage area of 3370 km^2.
- Mosul Dam: This was used to be referred to as the Saddam dam. It was constructed on the River Tigris 58 km upstream Mosul city. The maximum storage capacity is 14.45×10^9 m^3. The drainage area of the dam is 50,200 km^2. The total length of the reservoir is 75 km.
- Fars Dam (Bakhma): This dam is located on the Greater Zab River. The storage capacity reaches 17.1×10^9 m^3. The construction of the first stage of this dam is already finished but nothing was done after 1990 due to the political situation.
- Dokan Dam: This dam is located on the Lower Zab River 60 km west of Sulaymaniya city. Its storage capacity is 7.9×10^9 m^3. The area of the reservoir is 50 km^2.
- Derbendikhan Dam: This dam is located on the upper reaches of River Diyala with a total storage capacity of 3.0×10^9 m^3. The area of the reservoir is 121 km^2.

- Kut Barrage: This barrage was constructed on the River Tigris near Kut city to regulate the water. Two irrigation canals take water from this scheme (Gharaf, Dijalah).
- Duhok Dam: is located on Rubar River near Duhok city. The storage capacity of the dam is 50×106 m^3.
- Third River (Saddam River): This River is to collect saline water from irrigation projects on the Tigris and Euphrates Rivers. Large number of drainage canals supplies this river with water. At Nasiriyah it crosses the River Euphrates through a tunnel under the bed of the Euphrates River. It is about 550 km long. In its upper parts its discharge is 40 m^3/s and increase to 350 m^3/s in the lower reaches.
- Adhaim Dam: This dam was constructed on Adhaim River in 1999. Its storage capacity is 1.5×109 m^3. The Adhaim Dam project is located about 70 km upstream from the Adhaim's confluence with the Tigris. It diverts water at the confluence of the Adhaim and the Aq Su Rivers.
- Tharthar Scheme: Tharthar depression is located 65 km northwest Baghdad city between the Tigris and the Euphrates Rivers. The length and width of the depression is 100 km and 40 km respectively. Its total storage capacity is 85×109 m^3. This scheme is composed of:

Samara Regulating Barrage, which is a concrete structure located opposite Samara city. Intake of Ishaki irrigation canal is located upstream this regulating dam. The regulator barrage converts flood water to the Tharthar intake canal.

Protection Embankments, which extends to a distance of 62 km, and Intake Canal. The embankments are earth fill type for protection from floods upstream the Samara barrage. Flood water is taken through 36 gates. Each gate is 12 m wide. It can allow 9000 m^3/s when the water level upstream the Samara regulator is 69 m above sea level.

B. Proposed projects
- Upper Adhaim Dams: These are a number of small dams to be constructed on the tributaries of River Adhaim.
- Marhala Dam: To be constructed on the Taini tributary of the Lower Zab River.
- Razka Dam: To be constructed on the Razka valley to supply the Greater Zab River with water.
- Jumurka Dam: To be constructed on the Jumurka valley to supply the Greater Zab River with water.

3.3.2. Irrigation Projects on the River Euphrates
Dams:
- Qadisiyah Dam: It is located near Haditha city with a storage capacity of 9.98×10^9 m^3 with 0.24×10^9 m^3 of dead storage.
- Habaniyah Lake: Located west the Euphrates River near Faluja. Its storage capacity is 3.3×10^9 m^3.
- Razaza Lake: It is located south of Habaniyahlake with a dead storage capacity of 40×10^9 m^3.

Dams constructed within the western desert:
- Rutbah dam: It is located on Horan valley 30 km from Rutbah city. Its storage capacity is 32×10^9 m^3.
- Waleed project: Two small dams constructed southwest of Rutbah.
- Small Rutbah dam: Located on one of Horan valley's tributaries.
- Aghri dam: Located on the Aghri valley.
- Ebliah dam: Located on the Ebliah valley northeast of Rutbah.
- Husainiyah dam: Located on the Husainiyah valley northeast of Rutbah.
- Sury dam: Located south of Rutbah.
- Shibiga dam: Located southwest of Rutbah.
- Waleg dam: Located on the Waley valley.
- Kaieda dam: Located on the Kaeida valley.
- Asak dam: Located on the Asak valley.
- Amura dam: Located on the Amura valley.
- Um Turfat dam: Located on the Um Turfat valley.
- Rahaliya dam: Located on the Rahaliya valley.
- Ruwdha dam: Located on the Ruwdha valley.
- Mahmaja dam: Located on the Mahmaja valley.
- Abla dam: Located on the Abla valley.
- Jasan dam: Located on the Jasan valley.
- Minyama dam: Located on the Minyama valley.

Regulating schemes
- Ramadi barrage: Located near Ramadi. Its discharge capacity is 3600 m^3/s.

- Warar regulator: Its discharge capacity 2800 m³/s.
- Majara regulator: Its discharge capacity 850 m³/s.
- Dhiban regulator: Its discharge capacity 800 m³/s.
- Faluja regulator: Its discharge capacity 3600 m³/s.
- Hindiyah regulator: Its discharge capacity 2500 m³/s.
- Shatt Alhila regulator: Its discharge capacity 326 m³/s.
- Kufa regulator: Its discharge capacity 1400 m³/s.
- Mushkhab regulator: Its discharge capacity 750 m³/s.
- Abasiya regulator: Its discharge capacity 1200 m³/s.
- Shamiya regulator: Its discharge capacity 1100 m³/s.
- Ghalyun regulator: Its discharge capacity 250 m³/s.

4. The Problem

Scarcity of water resources in Iraq can be attributed to two main reasons:
A) External
B) Internal

4.1. External

4.1.1. Climate Change

MENA region with its arid climate is expected to be the most vulnerable in the world to the potential impacts of climate change [33]-[35]. The region is expected to suffer from higher temperatures and intense heat waves effecting inhabitants and the crops. Recent work for long term predictions of temperature and rainfall had indicated that the former will be increasing in Iraq while the latter will be decreasing (**Figure 8** and **Figure 9**) [17] [36]-[40].

IPCC [34] observed that there are different impacts on the physical and biological systems due to climatic change. Higher air temperatures will affect marine ecosystems and fisheries, less but more intense rainfall, causing both more droughts and greater flooding, sea level rise, more intense cyclones, new areas exposed to dengue, malaria, and other vector and waterborne diseases [1]. New record high had been noticed in some MENA countries including Iraq (52.0°C) [41].

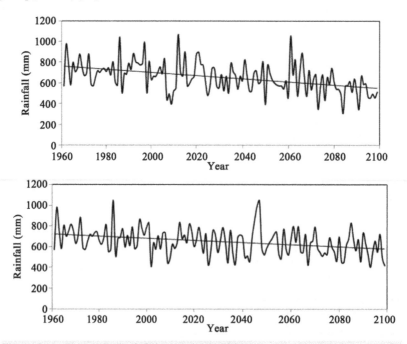

Figure 8. Average annual rainfall at northeast of Iraq for A2 scenario (upper) and B2 scenario (lower) compared with control period [36].

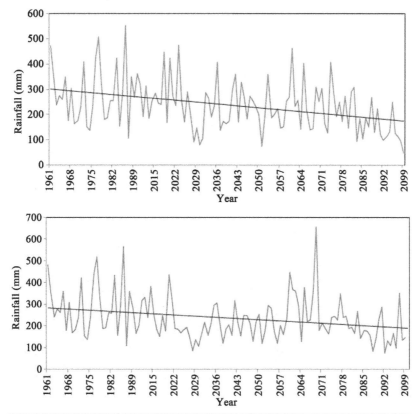

Figure 9. Average annual rainfall northwest Iraq for A2 scenario (upper) and B2 scenario (lower) compared with control period [37].

As a consequence of global warming, the size of oceans had expanded and sea level rise increased from 1.8 mm annually (1961-2003) to 3.1 mm (1993-2003) [34] [42]. AFED [33] and Dasgupta *et al.* [43], reported that the Gulf at its northern tip north of Kuwait and south of Iraq (Shatt el Arab) will be effected by sea level rise. Despite the limited coastline Iraq has on the Gulf region, the vulnerable low land areas extend as far inland as near Baghdad. In addition, Arnell [44] indicated that the flow of rivers in the Middle East will be decreased due to climatic changes at the end of the 21st century.

In MENA region will be warmer, drier and more variable where the temperatures are likely to rise 0.3°C or 0.4°C per decade. This is 1.5 times faster than the global average. Most of North Africa and the eastern Mediterranean will become drier [41]. Water supply and agriculture will be effected [45] [46]. Iraq will be greatly affected due to the fact that its one-third of cereal production (wheat and barley) is produced under rain-fed conditions at north of Iraq [47] [48]. Droughts and floods will increase as a result of increasing temperature [49]-[52]. This will affect the agriculture [46].

Dust storms are expected to be more frequent [33] [53] [54]. In addition, the precipitation is to decrease about 20% with temperature increase of 3°C to 5°C [55] (Elasha, 2010) and runoff will be reduced by 20% to 30% in most of MENA region by 2050 [56] (Milly *et al.*, 2005).

4.1.2. Shared Catchments

The Euphrates, Tigris and Shatt Al-Arab Rivers catchments are shared between the countries neighboring Iraq as shown in **Figure 10** and **Table 8** and **Table 9**. Due to the fact that all the major rivers in Iraq are shared with neighboring countries several agreements were signed with neighboring countries [2] but none of them had been put into practice (see as an example Burleson [57]).

During the 1970s Syria and Turkey started to construct dams on the Euphrates and Tigris Rivers which caused a major decrease in the flow of the rivers [1] [22] as well as deterioration of the quality of their water [58].

The situation became very tense both in 1975 and 1998 and war was narrowly averted due to water shortages after building dams [61].

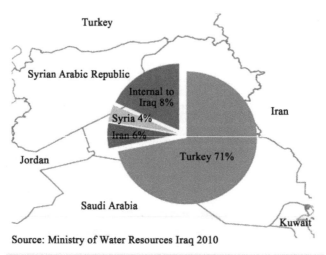

Source: Ministry of Water Resources Iraq 2010

Figure 10. Source of water for the Tigris and Euphrates Rivers [59].

Table 8. Rivers of Iraq and riparian countries.

River	Riparian Countries	Main Shared Tributaries
Euphrates	Iraq, Jordan, Saudi Arabia, Syria, Turkey	Sajur Jallab/Balikh Khabour
Tigris	Iraq, Syria, Turkey, Iran	FeeshKhabour Greater Zab Lesser Zab Diyala
Shatt Al-Arab	Iraq, Iran	Tigris Euphrates Karun Karkheh

Table 9. Contribution of countries to Euphrates and Tigris Rivers [60].

Tigris and Euphrates Rivers		Turkey	Iraq	Syria	Iran	Total
Discharge	%	78.1	8.1	0.5	13.3	
	Billion m³/year	65.7	6.8		0.5	84.2
Drainage area	%	20.5	46.0	9.0	19.0	
	Billion m³/year	170,000	469,000	77,000	37,000	819,000
River length	%	33.5	51.0	15.5	-	
	km	1630	2478	754	-	4862

Turkish Water Projects

In 1977, the Turkish government set a huge project referred to as Southeastern Anatolia Project (GAP) [62] (GAP, 2006). The component of the project includes 22 dams and 19 hydraulic power plants (**Table 10**) which are supposed to irrigate 17,000 km² of land [63] (Unver, 1997). The project is supposed to develop the southeastern provinces which cover 9.7% of the total area of Turkey which forms 20% of the agricultural land of the country. The overall volume of water to be captured is about100 km³ (while the required water to irrigate the supposed area is about 29 km³) which is three times more than the overall capacity of Iraq and Syrian reservoirs. Despite the continuous claims of the Turkish Government that GAP is purely development project, it seems that there are number of internal and external goals involved [64]-[69]. When GAP project is completed, then 80% of the Euphrates water will be controlled by Turkey [70]-[72].

Table 10. Dams of the GAP project in Turkey.

River Basin	Name of the Dam	Year of Completion
Euphrates	Ataturk	1992
	Birecik	2000
	Camgazi	1998
	Hancagrz	1988
	Karakaya	1987
	Karkamis	1999
	Buykcay	Suggested
	Catallepe	Suggested
	Gomikan	Suggested
	Kahta	Suggested
	Kayacik	Suggested
	Kemlin	Suggested
	Koeali	Suggested
	Sirmtas	Suggested
Tigris	Batman	1998
	Dicle	1997
	Kralkizi	1997
	Cizre	Suggested
	Garzan	Suggested
	Kayser	Suggested
	Ilisu	Under construction
	Silvan	Suggested

When Ilisu dam on Tigris River is operating then, Iraq will receive only 9.7 km^3 [73]. This implies that 47% of the river flow will be depleted. This in turn means that 696,000 ha of agricultural land will be abandoned due to water scarcity [1] [22]. Recent reports state that Tigris and Euphrates rivers will be completely dry by 2040 [20].

Syrian Water Projects

Syria built three main dams (**Table 11**) along Euphrates River with a total storage capacity of 16.1 km^3 for irrigation and electricity generation.

Syria used to receive 21 km^3/year of the Euphrates water prior 1990 which dropped to 12 km^3 in 2000 onward (40% reduction). As far as Iraq is concerned, the volume of water received dropped from 29 km^3 before 1990 [74] to 4.4 km^3 (85% reduction) now. Due to this reduction in water shares, the agricultural used land in both countries had been reduced from 650,000 ha to 240,000 ha. In addition, the quality of water deteriorated due to back water irrigation directed toward the main channel in its upstream reaches [1] [22].

Syria is planning to double its irrigated area (740,000 ha). This will increase its water withdrawal from 5 km^3 to 9 km^3 [75] and will cause:

• Diminishing of water for agriculture.
• Land degradation due to expected high salinity.
• More drying of the marshes causing more ecological damage.
• Further deterioration of the already bad water quality of the Euphrates (TDS is 1800 mg/L now).
• Less hydropower generation.
• Rising the risk of regional conflict
• Demographical implications where farmers and fishermen will leave their homes.
• Lower groundwater levels

Iranian Water Projects

In addition to the above, Iran had recently diverted all perennial valleys running toward Iraq inside Iran. Furthermore, water of Karkha and Karun Rivers had been almost completely diverted inside the Iranian borders and no water is contributing to Shatt Al-Arab River from these tributaries (**Figure 4** and **Figure 6**).

Shatt Al-Arab River is formed after the confluence of Tigris and Euphrates Rivers at Qurnah in Iraq. Its total length is 192 km and its width is about 300 m near Qurnah and increases downstream to 700 m near Basra city and to about 850 m near its mouth at the gulf area. Karun and Karkha Rivers usually contributes 24. 5 and 5.8 billion cubic meters (BCM) annually respectively (**Figure 6**). This forms about 41% of the water of Shatt Al-Arab. The decrease of the water discharge of the Tigris and Euphrates Rivers and the diversion of the water of Karun and Karkha tributaries caused the salinity to increase to 2408 mg/l in 2011 [2].

Table 11. Dams of Euphrates River in Syria.

Dam	Storage Capacity (km^3)	Year of Operation
Forat	14.163	1978
Baath	0.9	1989
Teshreen	1.883	2000
Total	16.943	

In addition, Iran and Kuwait signed an agreement in which Iran committed to supplying Kuwait with drinking water for a period of 30 years at a cost of USD 2 billion. About 300 MCM of water from the Karkheh River is to be conveyed to Kuwait through a 540-km pipeline [2]. Furthermore, Haweizeh Marsh used to extend over a vast area of 300,000 ha (in average conditions). Iran constructed a levy running along the international border through the Haweizeh Marsh which significantly reduced freshwater contribution from the Karkheh River and further jeopardize the subsistence of this important ecosystem [2].

4.2. Internal

The population growth rate in Iraq is very high (3.6%) where it was 20.4 million in 1995 and reached 32 million in 2013 [2]. The population density however, varies from 5 to 170 in habitants/km^2 in western desertic and the central part of the country respectively [1] [22]. One third of the inhabitants are involved in agricultural activities [76]. For this reason, most of the water is consumed in agricultural practices (90%) followed by industry (6%) and domestic uses (4%) [6] [11] [75]. About 38.5 km^3 of water was used for agricultural purposes in 1990since the percentage of the agricultural land is 19% - 25% (8.2 - 11.5 million ha) of the total area of Iraq. Recently, it is believed that this amount had been decreased slightly to 85% [1] [22]. This amount is used for an area of 8.2 million ha, which forms 70% of the total cultivable area. About 40% - 50% of this area is irrigable, while the remainder is rain fed and only 7% is of the area is supplied by ground water. Considering the soil resources, about 6 million hectares are classified as excellent, good or moderately suitable for flood irrigation. The irrigation potential is 63%, 35% and 2% for the Tigris, Euphrates and Shat Al-Arab Rivers respectively. Irrigation consumptive use reached 39 km^3 in 1991 and in 2003/2004 it was 22 km^3 equivalents to 44 km^3 of water derived, assuming 50% irrigation efficiency. Real efficiency might be 25% - 35%. Existing data estimates that the contribution of the agricultural sector was only 5% of Gross Domestic Product (GDP) which is usually dominated by oil (more than 60%). About 20% of the labour force is engaged in agriculture [1] [22].

It should be mentioned however, that the demand of the industrial sector decreased with the progress idling of the industrial capacity. Hydropower use including the evaporation from reservoirs reaches 10/annum BCM. Potable water usage in Iraq is about 350 liters/capita/day for the urban areas [77] and it used to reach 100% and 54% the urban and rural areas in 1991. The situation deteriorated in both quantity and quality afterwards and 33% of the population do not have access to water and sanitation. Current estimates indicate that water supply to urban areas is 94% and in rural areas is 67% [76]. Water services are limited to few hours per day and its quality does not meet WHO standards or Iraqi national water quality standards.

The available water in Iraq is still more than it is compared to the neighbouring countries with exception of Turkey, where it reaches 75 BCM which is equivalent to 2400 m^3 per capita per year [20] [78].

To estimate water demand in Iraq, MWR [79] studied this problem so that it can sustain good water quality of the inhabitants. The main facts concerning the water situation in Iraq can be summarized as follow:

In 2011, the demand was about 11 MCM/day in 2011, while the domestic water shortage was 1.7 MCM/day [77]. The goal of the government is to ensure water supplies to 91% of the population by 2015 [20]. The scenarios given by IMMPW [77] are based on lower on going consumption level from 350 to 200 litter/capita/day and with expected population of more than 34 million in 2015, the potable water demand will range between 8 and 13 MCM/day (**Table 12**) depending on overall (treating, conveying and distribution) efficiency. An additional 5 BCM of water required in sanitation sector due to the fact that the infrastructure is out of service losses have increased. About 79% of the population has access to drinking water (92% in urban and 57% in rural areas) [80]. In addition, those having access to drinking water it takes about 21 minutes in urban areas (42 minutes in rural areas) to get to the source to bring water for 17% of all households. The same survey showed that 21% have no

Table 12. Expected population and potable water demand for different water shares and different distribution system efficiency.

Water Share (l/capita/day)	Year	Population (Million)	All Population Demand (MCM/day)	91% Population Demand (MCM/day)	Real Water Demand (MCM/day)			
					Efficiency 50%	Efficiency 60%	Efficiency 70%	Efficiency 80%
350	2012	32	11.2	10.192	20.384	16.98667	14.56	12.74
	2015	34.98	12.243	11.14113	22.28226	18.56855	15.9159	13.92641
250	2012	32	8	7.28	14.56	12.13333	10.4	9.1
	2015	34.98	8.745	7.95795	15.9159	13.26325	11.3685	9.947438
200	2012	32	6.4	5.824	11.648	9.706667	8.32	7.28
	2015	34.98	6.996	6.36636	12.73272	10.6106	9.0948	7.95795

access to drinking water, 16% have daily problems, 7% have weekly problems, 15% have less than weekly problems and only 41% have reliable source. The World Bank [75] stated that the efficiency of the distribution network is very poor (32%) and it is deteriorating with time. For this reason water allocation per capita is decreasing with time since 1980 [81]. Furthermore, the quality of drinking water does not meet WHO standards or Iraqi national water quality standards [82]. Leakage, in both potable water distribution and sewage systems causes high contamination. In view of this situation large number of the population are suffering from various disease [20] [80] [83] [84].

Estimates of the agricultural sector in Iraq indicates that cropped land is about 1.9 million ha in recent years out of 4 million ha arable lands. According to Iraqi Ministry of Planning (MoP), Iraq is planning to increase the agricultural area in 2017 cropped by wheat and barley by 14% and 21% respectively [85]. In case all arable land is cropped then the water requirement is 50 BCM, assuming good irrigation efficiency.

Salinity of the soil in Iraq was noticed long time ago since the Sumerians [86] [87]. For this reason the efforts were concentrated since the mid-seventies of the last century to construct drainage projects. The major aims of drainage projects are to prevent water logging, control salinity and acidity as well as increasing cultivable areas [88]. The problem of drainage water intensified with the expansion of drainage of irrigated lands in Turkey and Syria which caused deterioration of the water quality of the two rivers. The Turkish government is planning to 1.8 million ha of the land through the GAP project (of which 270,000 ha is currently operational) while the Syrian government is planning to irrigate 650,000 ha (of which 325,000 ha are already irrigated from the Euphrates River in 2000) [2].

The Iraqi government tried to reduce the salinity in the two rivers by constructing what is referred to as "the third river". This river is 565 km long and it flows from Baghdad to the Gulf via the KhorZubair Canal. It was completed in 1992 after 30 years of work. It collects drainage water from more than 1.5 million ha of land between the Euphrates and Tigris Rivers. It is to resolve the chronic salinity problem affecting farmland between the two rivers, by collecting saline drainage water and preventing it from flowing into the Euphrates. The canals discharge into southern Hammar Marsh that is fed by the Euphrates and Tigris. About 17 million tons of salt reportedly flowed into the Gulf through the Third River in 1995 [2].

The salinity or Total Dissolved Solids (TDS) in Euphrates River at the Syrian-Iraqi borders is 600 mg/l which is already higher that the recommended level for irrigation (**Figure 11**) and it increases to more than 1200 mg/l (minimum) downstream at Samawah [75] [77]. Tigris River is in better situation relative to the Euphrates River (**Figure 12**). TDS values of the Tigris water at the Turkish Iraqi border are 280 - 275 mg/l and it reaches more than 1800 mg/l in Basra (IMMPW, 2011). The situation might be worse on the tributaries where TDS values in the Diyala River reaches 3705 mg/l [84] (Jawaheri and Alsahmari, 2009).

Within Iraq, the source of most of the back irrigation water is from irrigation projects (1.5 million ha) that are located in the central and southern parts of the country. Back irrigation water from these projects is directed to the main outfall drain which drains to the gulf in order to reduce the soil salinity [89]. Despite the presence of these drainage measures, the salinity increases downstream along the courses of the two rivers in conjunction with decreases in their discharges (**Figure 13** and **Figure 14**). This situation overstressed the agricultural sector in the southern Iraq. Tracking the changes of Tigris River's salinity shows that significant increase starts from Baghdad downward. This is due to the effect of the feedback from Tharthar depression toward Tigris River [90]. Where some of the Tigris flood flows is diverted at Samara Barrage to Tharthar depression which is highly saline (**Figure 15**), and then it is redirected for use in the river system with the salt washed from the depression.

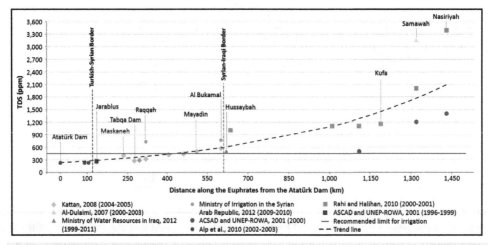

Figure 11. Salinity variation along Euphrates River since 1996 [2].

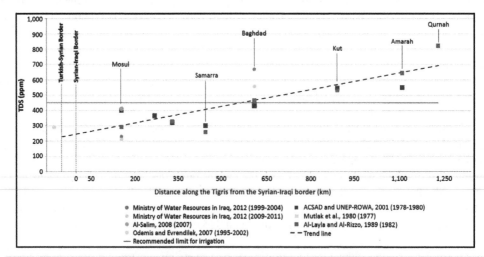

Figure 12. Salinity variation along Tigris River before 1983 and after 1995 [2].

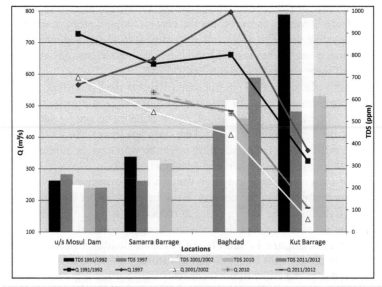

Figure 13. Variation of discharges (Q) and Total Dissolved Solids (TDS) along Tigris River inside Iraq.

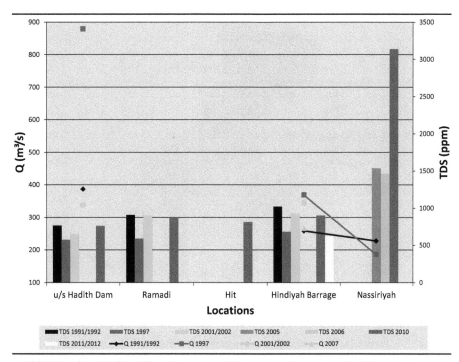

Figure 14. Variation of discharges (Q) and Total Dissolved Solids (TDS) along Euphrates River inside Iraq.

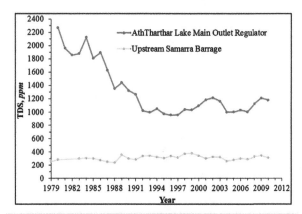

Figure 15. Total Dissolved Solids concentration at Samarra Barrage (the supplying point to Tharthar depression).

Recent estimates indicate that 4% of irrigated areas are severely saline, 50% are of medium salinity and 20% are slightly saline [91]. This forced the government to undertake a land rehabilitation program and a total of 700,000 ha were reclaimed. Later the situation deteriorated where recent estimates indicatethat4%of irrigated areas are severely saline, 50% are of medium salinity and 20% ares lightly saline [87].

In addition to the above mentioned problems, declining water flow of the Tigris and Euphrates Rivers, the repeated frequency of drought [92], water quality degradation and increasing soil salinity mean that large areas of Iraq are facing serious problems of desertification. It is believed that at least 45% to 75% of the area of Iraq has been substantially affected by desertification [93] [94]. In addition, during the Gulf wars, huge number of palm trees were destroyed which originally were acting as natural barriers against the expansion of desertification. In view of the above, a large number of farmers and fishermen left their land and villages were deserted. The expansion of desert areas led to frequent sand or dust storms [54] [95]. Between 2007 and 2009, 40% of cropland area experienced reduced crop coverage and 20,000 rural inhabitants left their homes (**Figure 16** and **Figure 17**) [82].

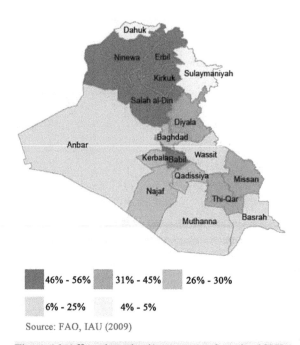

Source: FAO, IAU (2009)

Figure 16. Affected cropland/percentage of cropland [82].

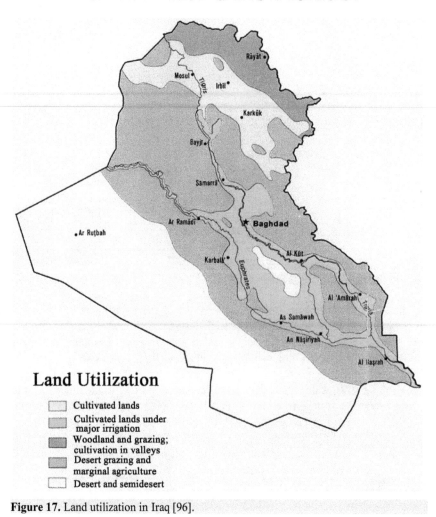

Figure 17. Land utilization in Iraq [96].

In view of the above, a large number of farmers and fishermen left their land and villages were deserted [37]. Between 2007 and 2009, 40% of cropland area experienced reduced crop coverage and 20,000 rural inhabitants left their homes [82].

The industrial sector in Iraq deteriorated greatly after the UN sanctions in the 1990s. All most all the factories are not functioning. This includes paper and petrochemical industries in Basra, textile industries in Hilla and Wasit, food and vegetable oil industries in Baghdad and Karbala, pharmaceutical industries in Samara, construction industries Anbar and Najaf. No documents found stating the quantity of water consumed by these industries. Iraqi government is looking for external investments for rehabilitation of these industries [97]. In addition, the electricity generation in Iraq is facing great demand to increase the production by 24% in 2017 [85]. This means more installation of thermal power plants, which are the popular electricity sources in Iraq, to replenish electricity shortage which will induce s more water demand.

In addition to the above, the marshes area requires rehabilitation after it was dried during the past regime. Marsh lands, which are known as the Garden of Eden, cover an area about 15,000 - 20,000 km^2 in the lower part of the Mesopotamian basin. Many agricultural projects were established on the dried lands at that time covering 1920 km^2 distributed over three provinces. After the fall of the previous in 2003 and with the help of other countries and international organizations, a new program started to restore the Iraqi marshes [21]. It is believed that 70% - 75% of the original areas of the marshes can be restored. Restoring the marshes requires 13 BCM without improving the quality of water in the marshes and extra 5 BCM to improve it [75].

In summary, the overall water required to be 75 to 81 BCM [75]. In 2010, UN [20], estimated the overall water demand excluding restoring the marshes is about 73 BCM and the available water is about 59 - 75 BCM. If the situation remains as it is, the Iraqi water supplies will drop to 43 BCM by 2015 and to 17.61 BCM in 2025 and the demand is 77 BCM or 66.85 BCM at the least. According to the World Bank [64], the Iraqi water deficit in 2030 will reach 25.55 BCM (37%) where the expected supply is 44 BCM only.

5. Conclusions and Recommendations

Iraq is facing water scarcity problem due to external and internal factors. External factors cannot be solved independently or in short term actions or planning like global climatic change and abusive water policies by riparian countries. In addition, these themes are to be addressed with regional and international cooperation.

In this context, there were number of agreements signed by riparian countries (especially Turkey, Syria and Iraq) for the water allocated. These agreements are well documented by ESCWA [2] and none of these had been put into practice for a long period of time. This is due to the fact that Turkey, considers the Euphrates and Tigris as Transboundary Rivers that falls under Turkish sovereignty as long as it is within its territory. Based on this, Turkey regards that it is not possible to share a commodity that constantly changes in quantity and quality, and in time and space due to the variable conditions of the hydrological cycle. While from Iraq and Syria point of view, these rivers are "international rivers" that should be treated as a shared entity by all riparian countries.

Long term expectations of the decrease of rainfall and temperature increase due to global climate change will inevitably reduce the quantity of the internal water resources and increase the desertification in Iraq which already reached 75%. Turkish water projects (GAP, Ilisu…etc.) will control 80% of the Euphrates water and 47% of Tigris River flow to Iraq. At least, 696,000 ha of agricultural land will be abandoned influenced by these projects. The diversion of the water of Karun and Karkha tributaries inside the Iranian borders caused very high increase of the salinity in Shatt Al-Arab. It is believed these problems are to be solved by discussion and cooperation between riparian countries under the UN umbrella so that it takes a formal international cover. In addition, facing the expected climate change cannot be achieved unless the countries of the region act collectively.

About 85% of water withdrawal water in Iraq is consumed for agriculture. Where, only 1.9 million ha out of 4 million ha of arable land are cultivated in recent years. Even with the degradation in the productivity of the industrial sector, the hydropower consumption including the evaporation from reservoirs reaches 10 BCM/annum. A plan to reduce the domestic consumption from 350 to 200 litre/capita/day is proposed parallel with other plans to supply potable water to 91% of the population by 2015.

Internal problems and related issues of water scarcity can be solved independently in relatively short period of time. These are related to mismanagement of water resources inside Iraq, such as water losses in the distribution networks, overuse of water by inefficient irrigation systems, pollute water resources by sewage feedback, increase water salinity…etc. Iraqi government is to adopt water demand management instead of water supply management policy.

Iraq suffers from many problems in its infrastructures, such as those related to water losses through its water distribution networks, water overuse in old irrigation schemes, and pollution of fresh water sources by back water from irrigation and sanitation. The efficiency of the distribution network is very poor (32%) and it is deteriorating with time. Quality of drinking water does not meet WHO standards or Iraqi national water quality standards and the high contaminated leaked sewage water threatens potable water networks. The estimated effluent that discharged untreated directly to the rivers is over 0.5 MCM/day.

The expansion of drainage of irrigated lands in Turkey and Syria will cause a further deterioration in the water quality of the rivers. The TDS level of Euphrates River at Syrian-Iraqi borders is 600 mg/l which is already higher that the recommended level for irrigation and increases to more than 1200 mg/l (as minimum) downstream at Samawah. Tigris River is relatively better than Euphrates at the borders, but the salinity increases significantly starting from Baghdad downstream by the influence of the feedback from Tharthar depression which is highly saline. The situation is worse on Diyala River where its TDS level reaches 3705 mg/l.

It had been noticed by various researchers and organizations that the problem is becoming more alarming with time where the gap between supply and demand is increasing. The supply of Tigris and Euphrates Rivers will be 43 and 17.61 BCM in 2015 and 2025 respectively while the demand is estimated to be between 66.8 to 77 BCM respectively. In addition to all of this, it had been reported that Tigris and Euphrates discharges will continue to decrease with time and they will be completely dry by 2040.

The following recommendations are believed to be able to help to overcome the water shortage problem in Iraq:

A) Strategic Water Management Plan

There is a great need for an integrated long term "National Water Master Plan" being designed and put in practice immediately. This plan should be a long term plan for at least 30 years. All authorities concerned should participate (e.g. Ministry of Water Resources, Ministry of Municipality and Public Work, Ministry of Agriculture, Water Resources staff at Universities, private sector, NGO's and representatives of regional and International organizations concerned) in this national plan. Water demand management is believed to be the backbone of such a plan. It should also consider the following:

- Rehabilitation of infrastructure which should cover water treatment plants, power plants as well as pumping stations.
- Public awareness program is vital so that all the people appreciate the seriousness of the problem they are facing.
- Defining institutional agenda that includes employment and training.
- Supply and demand should be considered. In this context new non-conventional water resources (water harvesting, treated waste water) are to be used.
- New irrigation techniques should be a priority also (e.g. sprinkler and drip irrigation).
- Private sector is to be enhanced to be involved in the investment.
- Inter-ministerial coordination is very important. This will save time, effort and money. More decentralization including budget in irrigation, water supply and sanitation sectors are to be practiced.

B) Irrigation and Agriculture

- The most efficient irrigation techniques that is suitable for the local conditions of soil, water availability and quality, crops … etc. should be considered. Traditional irrigation techniques should be abandoned because they cause waste of water. Drip irrigation is convenient for orchards using salty water while sprinkler irrigation is suitable for grains and both of them are more conservative than surface irrigation.
- Maintaining and developing the conveying systems to reduce the losses and increase conveying efficiency. Closed conduits are to be considered as conveying system that reduces evaporation losses and infiltration losses. It is also conservative in land use and protects irrigation water from contact with saline water table.
- Improving the drainage systems of cultivated lands to improve soil leaching and reduce soil salinity. Also the most effective modern drainage techniques such as perforated pipe drainage system in collecting and FITO treatment in treating drainage water should be considered. Return of drainage water to the rivers should be avoided and drainage projects are to be implemented (like the main outfall drain in the areas laying outside the service zone of this project).
- Reducing the use of chemical fertilizers and pesticides that can decrease the water quality when back irrigation water discharges to the rivers.
- Using FITO treatment with drainage water and sewage water to reuse it in restoring the marshes as well as the available fresh water.

- Institutions should reflect decentralization, autonomy and farmer empowerment.
- Enhance private investment in the agricultural sector.
- Public awareness program for farmers to use new suitable techniques in irrigation (drip irrigation and sprinkler irrigation).
- Partially built dams should be completed and measure is to be taken to build the suggested dams and irrigation projects. This will increase the storage capacity of dams about 27 km^3.

C) Water Supply and Sanitation

- Efficiency of distribution networks of drinking water specially diversion and supply down to the point of use which is most cost effective should be improved.
- Sewage networks leakages should be repaired and their efficiencies improved to prevent any source of pollution from these networks.
- New efficient projects should be put in practice to prevent water losses and pollution.
- Operation and maintenance activities are to be improved e.g. using ICT.
- New sewerage systems are to be implemented in areas that are not serviced. The sewage water is to be conveyed to the sewage treatment plants to reduce the pollution of groundwater from the leakage from old septic tanks.
- Installing new sewage treatment plants to satisfy the increased consumption of domestic sector. Membrane bioreactor technology can be used in these new treatment plants to reuse the treated water.

D) Research and Development

- A comprehensive data bank should be established which includes reliable climatological, hydrological, geological, environmental and soil data to be used by researchers and decision makers.
- Conducting research to import new suitable technologies in water resources and agriculture which suites Iraq environment.
- Non-conventional methods to augments water recourses are to be used. We believe that water harvesting techniques can be very effective and are relative cheap cost wise as well as treated waste water which can be used for special kind of plants.
- Training programs for technicians, engineers and decision makers about up to date technologies are very important.
- Execute pioneer projects which help in augmenting water resources, developing land productivity, minimizing water use and consumption.
- Setting the outlines of public awareness programs both for water use and agricultural activities. Such a program should start at primary schools onward.
- Universities and institutes should set special courses in arid region hydrology.
- Awarding of prizes for new innovations, pioneer researches and smart ideas in water resources and their management.
- Groundwater resources are still not exhausted, big efforts should spend to manage prudent using of this source and protect it from all kinds of pollution.

E) Regional Cooperation and Coordination:

- Defining institutional and technical needs for such cooperation.
- Cooperation on trans-boundary resources. Iraq, Turkey, Iran and Syria are to coordinate their efforts to reach reasonable agreements with riparian countries on water quotas under the UN umbrella.
- UN organizations (e.g. UNEP, UNDP, UNESCO etc.) and International institutions and organizations (FAO, WMO etc.) and universities should be asked to give their experience in this matter.
- Cooperation with other countries, organizations and companies in developed countries to help in giving advice for successful patterns of water management to get benefit from their experiences.

Acknowledgements

The research presented has been financially supported by Luleå University of Technology, Sweden and by "Swedish Hydropower Centre—SVC" established by the Swedish Energy Agency, Elforsk and Svenska Kraftnät together with Luleå University of Technology, The Royal Institute of Technology, Chalmers University of Technology and Uppsala University. Their support is highly appreciated. The Iraqi Ministry of Water Resources gratefully helped the authors to conduct this research.

References

[1]   Al-Ansari, N.A. (2013) Management of Water Resources in Iraq: Perspectives and Prognoses. *Journal of Engineering*, **5**, 667-668. http://dx.doi.org/10.4236/eng.2013.58080

[2]   ESCWA (Economic and Social Commission for Western Asia) (2013) Inventory of Shared Water Resources in Western Asia. SalimDabbous Printing Co., Beirut, 626 p.

[3]   Roger, P. and Lydon, P. (1994) Water in the Arab World. Harvard University Press, Massachusetts.

[4]   Biswas, A.K. (1994) International Waters of the Middle East—From Euphrates, Tigris to Nile. Oxford University Press, Bombay.

[5]   Allan, T. (2001) The Middle East Water Question. I.B.Tauris Publishers, London.

[6]   Al-Ansari, N. (1998) Water Resources in the Arab Countries: Problems and Possible Solutions. *UNESCO International Conference on World Water Resources at the Beginning of the 21st Century*, Paris, 3-6 June 1998, 367-376.

[7]   Cherfane, C.C. and Kim, S.E. (2012) Arab Region and Western Asia, UNESCWA. In: *Managing Water under Uncertainty and Risk, UN World Water Development Report* 4, *Chapter 33*.

[8]   Barr, J., Grego, S., Hassan, E., Niasse, M., Rast, W. and Talafré, J. (2012) Regional Challenges, Global Impacts. In: *Managing Water under Uncertainty and Risk, UN World Water Development Report* 4, *Chapter 7*.

[9]   Perry, C.J. and Bucknall, J. (2009) Water Resource Assessment in the Arab World: New Analytical Tools for New Challenges. In: Jagannathan, N.J., Mohamed, A.S. and Kremer, A., Eds., *Water in the Arab World: Managment Perspectives and Innovations*, World Bank, Middle East and North Africa Region, Washington, The World Express, 97-118.

[10] Hiniker, M. (1999) Sustainable Solutions to Water Conflicts in the Jordan Valley. Green Cross Internation, Geneva.

[11] Sadik, A.K. and Barghouti, S. (1994) The Water Problems of the Arab World: Management of Scarce Water Resources. In: Rogers, P. and Lydon, P., Eds., *Water in the Arab World*, Harvard University Press, Cambridge, Massachusetts, 4-37.

[12] Postel, S. (1992) Last Oasis: Facing Water Scarcity. Worldwatch Institute WW Norton & Co, New York.

[13] Charrier, B. and Curtin, F. (2007) A Vital Paradigm Shift to Maintain Habitability in the Middle East: The Integrated Management of International Watercourses. In: *Water for Peace in the Middle East and Southern Africa*, Green Cross International, Geneva, 11-1.

[14] Bazzaz, F. (1993) Global Climatic Changes and Its Consequences for Water Availability in the Arab World. In: Roger, R. and Lydon, P., Eds., *Water in the Arab Word: Perspectives and Prognoses*, Harvard University, Cambridge, 243-252.

[15] Al-Ansari, N.A., Salameh, E. and Al-Omari, I. (1999) Analysis of Rainfall in the Badia Region, Jordan. Al al-Bayt University Research Paper No.1, Mafraq, 66 p.

[16] Hamdy, A. (2013) Water Crisis and Food Security in the Arab World: The Future Challenges. http://gwpmed.org/files/IWRM-Libya/Atef%20Hamdy%20AWC.pdf

[17] Chenoweth, J., Hadjinicolaou, P., Bruggeman, A., Lelieveld, J., Levin, Z., Lange, M., Xoplaki, E. and Hadjikakou, M. (2011) Impact of Climate Change on the Water Resources of the Eastern Mediterranean and Middle East Region: Modeled 21st Century Changes and Implications. *Water Resources Research*, **47**, W06506, 1-18.

[18] Voss, K., Famiglietti, J., Lo, M., de Linage, C., Rodell, M. and Swenson, S. (2013) Groundwater Depletion in the Middle East from GRACE with Implications for Transboundary Water Management in the Tigris-Euphrates-Western Iran Region. *Water Resources Research*, **49**, 904-914. http://dx.doi.org/10.1002/wrcr.20078

[19] Nimah, M.N. (2008) Water Resources (2008) Report of the Arab Forum for Environment and Development. In: Tolba, M.K. and Saab, N.W., Eds., *Arab Environment and Future Challenges*, Chapter 5, 63-74, Arab Forum for Environment and Development, Beirut, Lebanon.

[20] UN (2010) Water Resources Management White Paper. United Nations Assistance Mission for Iraq, United Nations Country Team in Iraq, 20 p. http://iq.one.un.org/documents/100/white%20paper-eng_Small.pdf

[21] Al-Ansari, N.A., Knutsson, S. and Ali, A. (2012) Restoring the Garden of Eden, Iraq. *Journal of Earth Sciences and Geotechnical Engineering*, **2**, 53-88.

[22] Al-Ansari, N.A. and Knutsson, S. (2011) Toward Prudent Management of Water Resources in Iraq. *Journal of Advanced Science and Engineering Research*, **1**, 53-67.

[23] Wikipedia (2014) Tigris-Euphrates River System. http://en.wikipedia.org/wiki/Tigris%E2%80%93Euphrates_river_system

[24] Al-Ansari, N.A. (1979) Principles of Hydrogeology. Coll. of Science Press, Baghdad (In Arabic).

[25] Al-Sayyab, A., Al-Ansari, N.A., Al-Rawi, D., Al-Jassim, J., Al-Omari, F. and Al-Shaikh, Z. (1982) Geology of Iraq. Mosul University Press, Iraq, 280 (In Arabic).

[26] Al-Ansari, N.A., Assaid, H.I. and Salim, V.N. (1981) Water Resources in Iraq. *Journal of the Geological Society*, **15**, 35-42.

[27] Al-Ansari, N.A., Jassim, M.D., Thijeel, A., Abbas, H., Jawad, S.B. and Ahmad, A. (1990) Possibility of Using Western Desert for Agricultural Purposes. Ministry of Agricultural and Irrigation, Internal Report, 14 p.

[28] Al-Ansari, N.A., Dhari, M., Assad, N., Al-Shama, A. and Al-Jubori, S. (1993) Agricultural Use of Deserts Using Ground Water. *Journal of Water Research*, **12**, 113-124.

[29] Al-Ansari, N.A. and Al-Shama, A. (1995) Long Term Utilization of Ground Water in Liefiya-Lusuf Area, Western Desert. *Iraqi Science Journal*, **36**, 282-313.

[30] Al-Ansari, N.A., Rushdi, B. and AL-Shamma, A. (1995) Utilization of Ground Water for Agricultural Activities, South Nakhaib Area, Western Desert (IRAQ). *International Conference on Water Resources Management in Arid Regions*, Sultanate of Oman, 450-456.

[31] Alsam, S., Jassim, S.Z. and Hanna, F. (1990) Water Balance of Iraq: Stage 2, Geological and Hydrogeological Conditions. Report, Ministry of Irrigation, Iraq.

[32] Krasny, J., Alsam, S. and Jassim, S.Z. (2006) Hydrogeology. In: *Geology of Iraq*, 1st Edition, Dolin, Prague and Moravian Museum, Prague.

[33] AFED (Arab Forum for Environment and Development) (2009) Impact of Climate Change on Arab Countries. http://www.afedonline.org

[34] IPCC (Intergovernmental Panel on Climate Change) (2007) Climate Change 2007: Climate Change Impacts, Adaptation and Vulnerability. Cambridge University Press, Geneva.

[35] WRI (World Resources Institute) (2002) Drylands, People, and Ecosystem Goods and Services: A Web-Based Geo-Spatial Analysis. http://www.wri.org

[36] Al-Ansari, N.A., Abdellatif, M., Ezeelden, M., Ali, S. and Knutsson, S. (2014) Climate Change and Future Long Term Trends of Rainfall at North-eastern Part of Iraq. *Journal of Civil Engineering and Architecture*, **8**, 790-805.

[37] Al-Ansari, N.A., Abdellatif, M., Ali, S. and Knutsson, S. (2014) Long Term Effect of Climate Change on Rainfall in Northwest Iraq. *Central European Journal of Engineering*, **4**, 250-263.

[38] Al-Ansari, N.A., Abdellatif, M., Al-Khateeb, M. and Knutsson, S. (2014) Desertification and Future Rainfall Trends north Iraq. 8*th Edition of the International Scientific Congress of GIS and Geospace Applications Geotunis*, Tunis/Hammamet, 2-6 April 2014.

[39] Al-Ansari, N.A., Abdellatif, M., Zakaria, S., Mustafa, Y. and Knutsson, S. (2014) Future Prospects for Macro Rainwater Harvesting (RWH) Technique in North East Iraq. *Journal of Water Resource and Protection*, **6**, 403-420. http://dx.doi.org/10.4236/jwarp.2014.65041

[40] Zakaria, S., Al-Ansari, N. and Knutsson, S. (2013) Historical and Future Climatic Change Scenarios for Temperature and Rainfall for Iraq. *Journal of Civil Engineering and Architecture*, **7**, 1574-1594.

[41] World Bank (2011) DorteVerner and M. Fatma El-Mallah (League of Arab States), Adaptation to a Changing Climate in the Arab Countries. MNA Flagship Report, Sustainable Development Department, Middle East and North Africa Region.

[42] IPCC (Intergovernmental Panel on Climate Change) (2001) Climate Change 2001: The Scientific Basis. Contribution of Working Group I to the IPCC. Third Assessment Report 2001, Cambridge University Press, Cambridge.

[43] Dasgupta, S., Laplante, B., Meisner, C. and Yan, J. (2007) The Impact of Sea Level Rise on Developing Countries: A Comparative Study. World Bank Policy Research Working Paper 4136.

[44] Arnell, N.W. (2004) Climate Change and Global Water Resources: SRES Scenarios and Socio-Economic Scenarios. *Global Environmental Change*, **14**, 31-52. http://dx.doi.org/10.1016/j.gloenvcha.2003.10.006

[45] Oweis, T. and Hachum, A. (2004) Water Harvesting and Supplemental Irrigation for Improved Water Productivity of Dry Farming Systems in West Asia and North Africa. New Directions for a Diverse Planet. *Proceedings of the 4th International Crop Science Congress*, Brisbane, 26 September-1 October 2004. www.cropscience.org.au

[46] Medany, M. (2008) Impact of Climate Change on Arab Countries. Chapter 9. www.afedonline.org/afedreport/english/book9.pdf

[47] Food and Agriculture Organization (FAO) (2009) Irrigation in the Middle East Region in Figures— AQUASTAT Survey. 2008 Water Report 34, 2. http://www.fao.org/docrep/012/i0936e/i0936e00.htm

[48] Yahya, J.Q. (2002) Climate Change for the Semi-Mountainous Region in Iraq and Its Impact on the Productivity of Wheat and Barley. Unpublished Master Thesis, Department of Geography, University of Tikrit, College of Education,

Tikrit.

[49]   Wetherald, R.T. and Manabe, S. (2002) Simulation of Hydrologic Changes Associated with Global Warming. *Journal of Geophysical Research*, **107**, 4379. http://dx.doi.org/10.1029/2001JD001195

[50]   Karrou, M. (2002) Climatic Change and Drought Mitigation: Case of Morocco. *The First Technical Workshop of the Mediterranean, Component of CLIMAGRI Project on Climate Change and Agriculture*, Rome, 25-27 September 2002, 7p.
        http://www.semide.net/media_server/files/semide/topics/waterscarcity/background/climatic-change-and-drought-mitig
        ation-case-morocco/ws01_38.pdf

[51]   Abbas, A. (2002) Drought Suppression Procedures for Dry Lands. *The First Technical Workshop of the Mediterranean, Component of CLIMAGRI Project on Climate Change and Agriculture*, Rome, 25-27 September 2002, 12p.
        http://www.docstoc.com/docs/32615805/DROUGHT-SUPPRESSION-PROCEDURES-FOR-DRY-LANDS-EXAMP
        LE-SALAMIA

[52]   Mougou, R. and Mohsen, M. (2005) Hendi Zitoune Case Study: Agro Climatic Characterization and Evapotranspira-tion Evolution in Climate Change Conditions. Fourth Tunisian Semi-Annual Report. Contribution to the AIACC AF 90 North Africa Project: Assessment, Impacts, Adaptation, and Vulnerability to Climate Change on North Africa: Food and Water Resources.

[53]   AL-Bayati, F.F. (2011) Climatic Conditions and Their Impact on the Geographical Distribution of the Dust Storms Empirical Study at Al-Anbar Province-Iraq. Anbar University, *Journal for the Humanities*, **1**, 26-38.

[54]   Sissakian, V., Al-Ansari, N.A. and Knutsson, V. (2013) Sand and Dust Storm Events in Iraq. *Natural Science*, **5**, 1084-1094. http://dx.doi.org/10.4236/ns.2013.510133

[55]   Elasha, B.O. (2010) Mapping of Climatic Change Threats and Human Development Impacts in the MENA Region. United Nations Development Programme, MENA Human Development Report (AHDR), Research Paper Series 2010.
        http://www.MENA-hdr.org/publications/other/ahdrps/paper02-en.pdf

[56]   Milly, P.C.D., Dunne, K.A. and Vecchia, A.V. (2005) Global Patterns of Trends in Stream Flow and Water Availabili-ty in a Changing Climate. *Nature*, **438**, 347-350. http://dx.doi.org/10.1038/nature04312

[57]   Burleson, E. (2005) Equitable and Reasonable Use of Water within the Euphrates-Tigris River Basin. Environmental Law Institute, Washington.
        http://www.internationalwaterlaw.org/bibliography/articles/general/BurlesonTigris-Euphrates.pdf

[58]   Kamel, A., Sulaiman, S. and Mustaffa, S. (2013) Study of the Effects of Water Level Depression in Euphrates River on the Water Quality. *Journal of Civil Engineering and Architecture*, **7**, 238-247.

[59]   Ministry of Water Resources-Iraq (MWR) (2010) Annual Report. National Centre for Water Resources Management.

[60]   Biedler, M. (2004) Hydropolitics of the Tigris-Euphrates River Basin with Implications for the European Union. Re-search Paper no. 1, Centre European de Reserche International et Strategique.
        http://www.ceris.be/fileadmin/library/Research%20Papers/1%20Hydropolitics%20of%20the%20Tigris-Euphrates%20
        Rivers%20Bassin%20with%20implantations%20for%20the%20European%20Union.pdf

[61]   Alkanda, A., Freeman, S. and Placht, M. (2007) The Tigris-Euphrates River Basin: Mediating a Path towards Alnakhla. The Flecher School Journal for Issues Related to Southwest Asia and Islamic Civilization.
        http://fletcher.tufts.edu/AlNakhlah/Archives/~/media/Fletcher/Microsites/al%20Nakhlah/archives/pdfs/placht-2.pdf

[62]   GAP (2006) South-Eastern Anatolia Project: Latest Situation. http://www.gap.gov.tr/English/Genel/sdurum.pdf

[63]   Unver, I. (1997) Southeastern Anatolia Project (GAP). *International Journal of Water Resources Development*, **13**, 453-484. http://dx.doi.org/10.1080/07900629749575

[64]   Shams, S. (2006) Water Conflict between Iraq and Turkey. Middle East News. http://www.mokarabat.com/m1091.htm

[65]   Alnajaf News Net (2009) The GAP Project and Its Negative Implications on Iraq (in Arabic).
        http://www.arabo.com/links/,199,225,218,209,199,222/,197,218,225,199,227/45188.html

[66]   Waterbury, J. (1993) Transboudary Water and the Challenge of International Cooperation in the Middle East. In: Roger, R. and Lydon, P., Eds., *Water in the Arab Word: Perspectives and Prognoses*, Harvard University, Cambridge, 39-64.

[67]   Alsowdani, M. (2005) GAP Project and Its Economic Negative Implications on Syria and Iraq, in Al-Itehad News.
        http://www.alitthad.com/paper.php?name=News&file=print&sid=19030

[68]   National Defense Magazine (2009) Turkish Israeli Partnership in GAP Southeastern Anatolian Project. Official Site of the Lebanese Army. http://www.lebarmy.gov.lb/article.asp?ln=ar&id=2901

[69]   Murakami, M. (1995) Managing Water for Peace in the Middle East: Alternative Strategies. United Nations University Press, Tokyo. http://archive.unu.edu/unupress/unupbooks/80858e/80858E04.htm

[70]   Beaumont, P. (1995) Agricultural and Environmental Changes in the Upper Euphrates Catchment of Turkey and Syria

and Their Political and Economic Implications. *Applied Geography*, **16**, 137-157.

[71] Alyaseri, S. (2009) GAP Project: Dangerous Consequences on Life in Iraq. Official Site of Iraqi Council for Peace and Unity. http://www.marafea.org/paper.php?source=akbar&mlf=copy&sid=11556

[72] Robertson, C. (2009) Iraq Suffers as the Euphrates River Dwindles. The New York Times. July 13th, 2009. http://www.nytimes.com/2009/07/14/world/middleeast/14euphrates.html?_r=0

[73] Iraqi Civil Society Solidarity Initiative (2014) Ilisu Dam and Legal Considerations in Iraq. http://www.iraqicivilsociety.org/archives/2902

[74] Majeed, Y. (1993) The Central Regions: Problems and Perspectives. In: Roger, R. and Lydon, P., Eds., *Water in the Arab Word*: *Perspectives and Prognoses*, Harvard University, Cambridge, 101-120.

[75] World Bank (2006) Iraq: Country Water Resources, Assistance Strategy: Addressing Major Threats to People's Livelihoods. Report no. 36297-IQ, 97 p.

[76] World Bank (2014) Country at a Glance: Iraq. http://data.worldbank.org/country/iraq

[77] Iraqi Ministry of Municipalities and Public Work (IMMPW) (2011) Water Demand and Supply in Iraq: Vision, Approach and Efforts, GD for Water. http://www.mmpw.gov.iq/

[78] Stockholm International Water Institute (SIWI) (2009) Water Resources in the Middle East. Background Report to Seminar on Water and Energy Linkages in the Middle East, 9.

[79] Ministry of Water Resources-Iraq (MWR) (2014) The Strategic Study for Iraqi Water Resources and Lands, in Arabic, http://www.mowr.gov.iq/?q=node/902

[80] Multiple Indicator Cluster Survey (MICS) (2007) IRAQ: Monitoring the Situation of Children and Women. Final Report. http://www.childinfo.org/files/MICS3_Iraq_FinalReport_2006_eng.pdf

[81] World Bank (2013) Food and Agriculture Organization, AQUASTAT, Renewable Internal.

[82] IAU (Inter-Agency Information and Analysis Unit) (2011) Water in Iraq Factsheet. http://www.iauiraq.org/documents/1319/Water%20Fact%20Sheet%20March%202011.pdf

[83] A Climate for Change (2012) Water Supply and Sanitation in Iraq. http://www.aclimateforchange.org/profiles/blogs/water-supply-and-sanitation-in-iraq

[84] Jawaheri, E.A. and Alsahmari, R.A. (2009) Water Problems of Iraq and Possible Solutions. *Journal of Law and Politics*, **2**, 9-61.

[85] MoP (Ministry of Planning-Iraq) (2013) National Development Plan 2013-2017. Baghdad.

[86] Al-Ansari, N.A. (2005) Applied Surface Hydrology. Al al-Bayt University Publication, Mafraq (In Arabic).

[87] The Encyclopedia of Earth (2008) Water Profile of Iraq. http://www.eoearth.org/article/Water_profile_of_Iraq

[88] Ritzema, H. and Braun, H. (1994) Environmental Aspects of Drainage, Chapter 25. In: *Environmental Drainage Principles and Applications*, International Institute for Land Reclamation and Improvement (IILRI), Publication No. 16, 1041-1065.

[89] Merry, M. (1992) Iraq Builds "Third River" Project Despite No-Fly Zone and Embargo. *Executive Intelligence Review*, Washington, DC, **19**, 8-10.

[90] Al-Shamari, H.M. (1994) Hydro Chemical Investigations of Tharthar Lake and Its Possible Effect on the Hydrochemistry of Euphrates River. M. Sc. Thesis, University of Baghdad, Baghdad (in Arabic).

[91] Consulting Engineering Bureau (CEB) (2011) Lakes Testing Study. College of Engineering, University of Baghdad, Baghdad.

[92] Ali, S.M., Qutaiba, A.S., Hussan, M. and Al-Azawi, F.W. (2013) Fluctuating Rainfall as One of the Important Cause for Desertification in Iraq. *Journal of Environment and Earth Science*, **3**, 25-33.

[93] Raphaeli, N. (2009) Water Crisis in Iraq: The Growing Danger of Desertification, Investors Iraq, 23 July 2009, no. 537. http://www.investorsiraq.com/showthread.php?132306-Water-Crisis-in-Iraq-The-Growing-Danger-of-Desertification

[94] Al-Saidi, A. and Al-Juaiali, S. (2013) The Economic Costs and Consequences of Desertification in IRAQ. *Global Journal of Political Science and Administration*, **1**, 40-45.

[95] UN-Iraq (2011) Sand and Sand Storm Fact Sheet. http://reliefweb.int/sites/reliefweb.int/files/resources/SDS%20Fact%20Sheet.pdf

[96] Food and Agriculture Organization (FAO) (2005) Country Profile: Land Use and Water Resources. http://www.fao.org/ag/agp/AGPC/doc/Counprof/Iraq/Iraq.html

[97] Ministry of Industry and Minerals-Iraq (MIM) (2008) Industrial Investment Opportunities in IRAQ. Investment Department, Baghdad, no. 537. http://www.investorsiraq.com/showthread.php?132306-Water-Crisis-in-Iraq-The-Growing-Danger-of-Desertification

10

Analysis of Water Stress Prediction Quality as Influenced by the Number and Placement of Temporal Soil-Water Monitoring Sites

Luan Pan[1], Viacheslav I. Adamchuk[1*], Richard B. Ferguson[2],
Pierre R. L. Dutilleul[3], Shiv O. Prasher[1]

[1]Department of Bioresource Engineering, McGill University, Ste-Anne-de-Bellevue, Canada
[2]Department of Agronomy and Horticulture, University of Nebraska-Lincoln, Lincoln, USA
[3]Department of Plant Science, McGill University, Ste-Anne-de-Bellevue, Canada
Email: [*]viacheslav.adamchuk@mcgill.ca

Abstract

In an agricultural field, monitoring the temporal changes in soil conditions can be as important as understanding spatial heterogeneity when it comes to determining the locally-optimized application rates of key agricultural inputs. For example, the monitoring of soil water content is needed to decide on the amount and timing of irrigation. On-the-go soil sensing technology provides a way to rapidly obtain high-resolution, multiple data layers to reveal soil spatial variability, at a relatively low cost. To take advantage of this information, it is important to define the locations, which represent diversified field conditions, in terms of their potential to store and release soil water. Choosing the proper locations and the number of soil monitoring sites is not straightforward. In this project, sensor-based maps of soil apparent electrical conductivity and field elevation were produced for seven agricultural fields in Nebraska, USA. In one of these fields, an eight-node wireless sensor network was used to establish real-time relationships between these maps and the Water Stress Potential (WSP) estimated using soil matric potential measurements. The results were used to model hypothetical WSP maps in the remaining fields. Different placement schemes for temporal soil monitoring sites were evaluated in terms of their ability to predict the hypothetical WSP maps with a different range and magnitude of spatial variability. When a large number of monitoring sites were used, it was shown that the probability for uncertain model predictions was relatively low regardless of the site selection strategy. However, a small number of monitoring sites may be used to reveal the underlying relationship only if these locations are chosen carefully.

---

[*]Corresponding author.

Keywords

On-the-Go Soil Sensing, Variable-Rate Irrigation, Electrical Conductivity, Site-Specific Water Management, Soil Matric Potential

1. Introduction

When pursuing site-specific crop management, temporal variability in soil water content is frequently as important as spatial variability. Thus, in order to optimize irrigation water management, one should combine the knowledge of changes of soil water holding capacity across a field with temporal monitoring of the actual water content available to plants during the most critical phases of crop production. Implementing this "precision irrigation" strategy means optimizing both the quantity and the timing of irrigation that may vary across a field due to different soil and growing conditions. Using traditional soil analysis practices, without considering soil spatial heterogeneity, is not optimal when it comes to site-specific water management.

Proximal soil sensing technology makes it possible to obtain high-resolution maps pertaining to different soil properties at a relatively low cost [1]. Unfortunately, the relationships between the data detected on-the-go and agronomic soil parameters such as water content are frequently site-specific. In addition, the amount of water stored in the soil profile changes not only spatially, but also temporally. Therefore, sensor-based maps have been used to define the spatial variability of soil properties influencing water movement and storage across a landscape, and this information has been used to define relatively homogeneous management zones that have been evaluated separately [2].

Increasingly, wireless technology is used to achieve temporal monitoring of soil conditions. Such systems allow the producer to obtain information about soil water content, temperature, and other properties in real-time from a remote location [3]. This technology greatly improves the convenience of monitoring soil water for the producer. Irrigation system managers have employed the data to optimize the use of resources in response to dynamic changes in soil water content and to reduce the risk of crop water stress [4]-[6].

Selecting a number of strategic locations within the field is not trivial and is mostly subjective. Practitioners who use high-resolution data layers apply the following general rules: 1) cover the entire range of data from each source, 2) avoid field boundaries and other transition zones, and 3) spread locations over the entire field. Quality coverage of the entire data space is important to make sure that both dry and wet field locations are monitored. Boundary and transition areas are avoided to make sure monitoring locations and spatially variable field characteristics represent the same local conditions. Finally, the geographic spread is needed to account for any additional uncertainties, such as possible spatially variable amounts of rainfall.

While the criteria listed above are useful, they do not translate into an operational algorithm and, therefore, can produce numerous solutions with different degrees of satisfaction. In principle, this process is similar to prescribing targeted sampling locations to either calibrate high-resolution data, or to quantify the agronomic soil attributes of established management zones [7]-[11]. Determination of the number of target sampling locations is another critical task.

With an ultimate goal of developing an algorithm for optimization of a wireless sensor network design, the **objective** of this study was to quantify the influence of the number and locations of temporal soil-water monitoring sites on the quality of water stress potential predictions within a set of hypothetical fields representing different levels of spatial variability.

2. Materials and Methods

2.1. Experimental Data

Seven agricultural fields in Nebraska (**Figure 1**) were mapped using on-the-go sensing technology with a Veris® 3150 or 3100 unit (Veris Technologies, Inc., Salina, Kansas)[1] equipped with an RTK-level global navigation satellite system (GNSS) receiver. Apparent soil electrical conductivity (EC_a) and field elevation (altitude) were used to assess the spatial variability for soil water storage. In one of these fields (a 37-ha field at the Agricultural

---

[1]Mention of a trade name, proprietary product, or company name is for presentation clarity and does not imply endorsement by the authors, the University of Nebraska-Lincoln, or McGill University, nor does it imply exclusion of other products that may also be suitable.

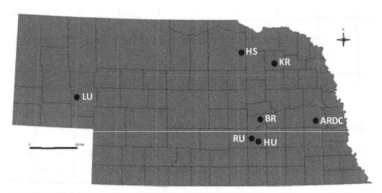

Figure 1. Research fields in Nebraska, USA.

Research and Development Center (ARDC) near Mead, Nebraska, USA), eight locations under a center-pivot irrigator were selected for monitoring the soil matric potential and the temperature using wireless sensor technology [12]. For each location, the Water Stress Potential (WSP), originally called Water Stress Index, was calculated as:

$$WSP = \sum_{i=1}^{4} w_i \cdot \left(\frac{\psi_i}{\psi_i'} - 1 \right) \tag{1}$$

where w_i is the weighting factor for water extracted in the plant root zone for depth increment i; Ψ_i is the soil matric potential measurement at the ith depth, kPa; and ψ_i' is the soil-specific soil matric potential at certain threshold level of soil water depletion (25% depletion was used in this research as half of the most frequently cited percent depletion causing water stress in corn) at the ith depth, kPa.

The following regression model [12] was used to define time-specific relationships between the high-density spatial data and WSP:

$$WSP = \beta_0 + \beta_1 \cdot EC_a + \beta_2 \cdot Elev + \beta_3 \cdot EC_a \cdot Elev \tag{2}$$

where EC_a is apparent soil electrical conductivity, mS/m; Elev is field elevation, m; $\beta_0, \beta_1, \beta_2, \beta_3$ are model coefficients.

Since a different set of regression coefficients was established for any point in time, it was possible to generate WSP maps using each new regression model. Thus, 99 sets of regression model coefficients were determined for every day of the 2009 growing season in the ARDC field (**Figure 2**). July 30th, 2009 (day 33) was chosen as an example of a regression equation that linked WSP, EC_a and field elevation in the middle of the growing season:

$$WSP = 139.52 - 27.82 \cdot EC_a - 0.40 \cdot Elev + 0.08 \cdot EC_a \cdot Elev \tag{3}$$

This equation can be generalized by using relative field elevations (obtained by subtracting the median elevation) instead of the original field elevation values:

$$WSP = -0.26 - 0.13 \cdot EC_a - 0.40 \cdot Elev_{rel} + 0.08 \cdot EC_a \cdot Elev_{rel} \tag{4}$$

where $Elev_{rel}$ is the relative field elevation.

To evaluate the quality of the WSP prediction using different distributions of temporal monitoring stations, it is important to know the WSP value in every location in the field, or at least, across a substantial number of validation locations. Unfortunately, this is not practical because of the high cost of temporal monitoring locations and numerous uncertainties at a fine scale. Therefore, an alternative analytical approach has been used in this study. Equation (4) was assumed to be valid for the other six fields (BR, HS, HU, KR, LU, and RU), which would allow modeling hypothetical maps of WSP at the same resolution as soil EC_a and field elevation maps. These new fields represented diversified conditions in terms of field topography and soil heterogeneity (**Table 1**). In each case, both EC_a and elevation data were interpolated (Inverse Distance Weighting) using a 10 × 10 grid, and corresponding maps were obtained for the calculated WSP values (**Figure 3**). Each of these maps is a hypothetical representation of what could be the WSP distribution across a given field at a specific time, assuming the regression Equation (4) has a perfect fit.

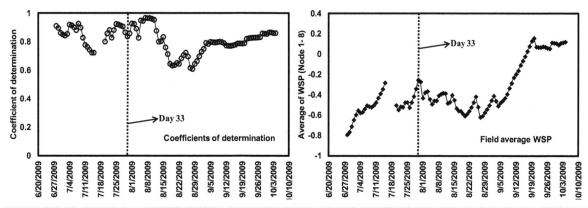

Figure 2. Coefficients of determination and field average WSP from Pan *et al.* [12].

Table 1. Field description.

Field ID	ARDC (Day 33)	BR	HS	HU	KR	LU	RU
Field area (ha)	37	27	47	72	57	49	49
Number of point measurements	8	7477	10,171	14,988	13,547	16,020	11,364
Field elevation (m)							
Average	351	517	583	563	507	1202	564
Standard Deviation	5.63	0.53	2.25	1.62	5.97	2.26	4.00
Min	342	516	577	559	499	1195	559
Max	359	519	589	568	529	1212	577
Range	17.0	3.4	11.8	8.4	29.9	16.8	18.1
Shallow (0 - 30 cm) soil EC_a (mS/m)							
Average	6.28	0.76	2.3	7.2	2.5	6.2	4.0
Standard Deviation	4.46	0.37	0.90	2.02	1.79	2.00	1.80
Min	0.68	0.22	0.7	2.4	0.4	1.6	0.2
Max	12.94	3.72	9.9	14.0	9.9	15.1	11.3
Range	12.27	3.50	9.2	11.6	9.5	13.4	11.1
Calculated *WSP*							
Average	−0.25	−0.37	−0.6	−1.3	−0.4	−1.0	−0.6
Standard Deviation	0.78	0.16	0.55	0.48	1.16	0.49	0.73
Min	−0.71	−1.05	−2.2	−3.7	−5.7	−4.3	−3.3
Max	1.61	−0.02	1.2	−0.1	4.8	1.3	4.0
Range	2.32	1.03	3.4	3.7	10.5	5.6	7.3

2.2. Error Simulation

Since in reality the regression model linking the *WSP* with EC_a and elevation is an approximation, the actual *WSP* map is different from the calculated *WSP*. To model a *WSP* map that could represent the actual state-of-nature, three different error simulation strategies were used to represent different degrees of spatial structure. **Table 2** summarizes the spatial structure used to simulate error surfaces. All three surfaces had the same total variance (sill = 0.04), which was similar to what had been observed in the ARDC field, and a mean error equal to zero, which was used to avoid bias between different modeled *WSP* maps (**Figure 4**). The first surface was assumed to have no spatial structure, while the second surface was simulated using an arbitrary spherical variogram model with a zero nugget effect and 300-m range of spatial dependency. The third surface was an intermediate case with a range two times smaller and a partial sill equal to the nugget effect.

Figure 5 illustrates three simulated *WSP* maps obtained for the HS field. Such error surfaces covering the extents of each field were superimposed on the calculated *WSP* maps. Then, these error surfaces were clipped to the

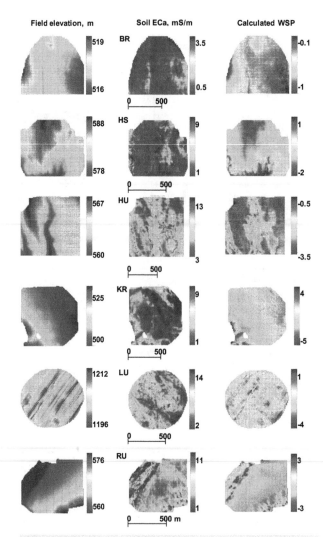

Figure 3. Maps of field elevation, soil EC_a, and calculated *WSP*.

Table 2. Variogram parameters for three error surfaces (ES 1-3).

Variogram Parameters	ES 1	ES 2	ES 3
Average	0	0	0
Nugget effect	0.04	0	0.02
Partial sill	0	0.04	0.02
Sill	0.04	0.04	0.04
Range (m)	0	300	150

actual field shape and re-scaled to maintain a zero-mean error and the simulated spatial structure. Each of the resulting maps was an example of the hypothetical WSP distribution across the landscape that could have actually taken place in a given field at a specific time. Predicting these maps with the least possible error would require an optimal network of temporal monitoring stations.

2.3. Soil-Water Monitoring Site Selection

In practice, the definition of the optimum-guided sampling scheme is rather vague. There are many parameters that can quantify 1) spatial separation, 2) spread across both sets of measurements (in this case, EC_a and field elevation), and 3) local homogeneity within each set of measurements. Furthermore, there are several different

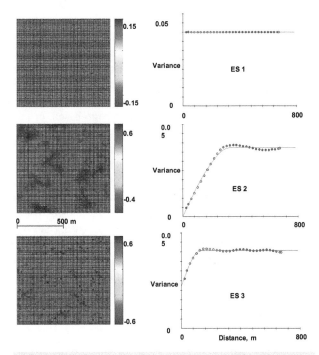

Figure 4. Maps and semivariograms of simulated error surfaces.

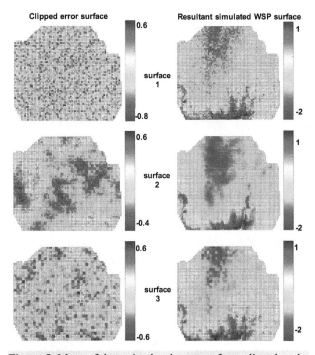

Figure 5. Maps of three simulated error surfaces clipped to the HS field shape and the resultant simulated WSP maps.

ways to derive the overall objective function as a combination of these parameters. In this study, the criteria used by Adamchuk *et al.* [11] were applied. These criteria included: 1) complete spatial field coverage using the S optimality criterion [13]; 2) even distributions throughout both data layers (in this case, EC_a and field elevation) using the D optimality criterion [13]; and 3) the relative homogeneity of the selected sites using the sum of squared differences between the measurements obtained in each location and those of the nearest neighbors. While S-optimality seeks to maximize the harmonic mean distance between each monitoring location in relation to all other monitoring locations, D-optimality increases with greater coverage of the range of soil EC_a and field elevation maps. The overall objective function was the geometric mean of these criteria normalized against the median of a large number of random selections. The optimized set of monitoring sites with the overall objective function closest to one was selected among 1000 randomly selected sets. In practice, the results of this optimization can be achieved through a hypercube sampling with an added requirement of geographic spread and a penalty for selecting monitoring sites in field locations with highly variable EC_a or field elevation. Although alternative site section algorithms could be acceptable, typically, they rely on either design-based, or model-based, inference [9]; in fact, both can be important.

To compare different site selection strategies, the performance of 100 randomly selected sets of soil-water monitoring sites were evaluated along with the optimally selected placements of these sites. This was repeated for each field and the number of potential soil-water monitoring locations varied from 1 to 10. **Figure 6** illustrates the relationship between EC_a and field elevation for the entire field, and indicates the best set of five (Set A) and the second-best set of five (Set B) soil-water monitoring locations chosen using a semi-automatic optimization process. Therefore, two optimized monitoring site selections with 5 (using Set A) and 10 (using Sets A & B) were assessed by comparing them to multiple random selections with the same number of monitoring sites.

Similar to the ARDC field, regression analysis was used to predict the simulated WSP values using EC_a and field elevation data that corresponded to locations dedicated to soil-water monitoring. This was done with Equation (4) modified for less than 4 soil-water monitoring sites as follows:

$$WSP = \begin{cases} \beta_0 + \beta_1 \cdot EC_a + \beta_2 \cdot Elev_{rel} + \beta_3 \cdot EC \cdot Elev_{rel} & \text{for } N \geq 4 \\ \beta_0 + \beta_1 \cdot EC_a + \beta_2 \cdot Elev_{rel} & \text{for } N = 3 \\ \beta_0 + \beta_1 \cdot EC_a & \text{for } N = 2 \\ \beta_0 & \text{for } N = 1 \end{cases} \tag{5}$$

It is necessary to mention that when N was equal to 2, either β_1 or β_2 could be left within the equation (depending on which data layer was more influential, EC_a or $Elev_{rel}$). In this study, a change of EC_a appeared to have more effect on *WSP* than field elevation. These regression equations were applied to the six test fields to produce predicted *WSP* maps. The Mean Squared Error (MSE) between *WSP* predictions and modeled *WSP* values was assumed to be a primary indicator of water stress prediction quality. Finally, the percentage of randomly selected sets of soil-water monitoring locations with MSE higher than the MSE calculated for the optimized set of monitoring locations provided an estimated probability of acceptable quality of water stress prediction using a random selection of monitoring sites.

3. Results and Discussion

Table 3 summarizes MSE estimates for each field. The results, based on the WSP surface with random error (ES 1) and with spatially distributed error (ES 2), are shown in **Figure 7**. The main difference is that a smaller number of soil-water monitoring sites resulted in a higher MSE. This increase was moderate for the most successful random monitoring site selection (min MSE) and for the optimized selection, but the chance of the random site selection producing a high MSE increased with a decreasing number of sites. Based on **Table 4**, simulating WSP using a random error (ES 1) resulted in 87% - 100% of random site selections producing MSE values greater than for the five-site optimized selection. However, this number ranged between 48% and 83% when the proportion of monitoring locations was doubled. This means that the use of a large number of sites reduces the chance of randomly selecting the wrong soil-water monitoring locations. However, the relatively high cost of equipment does not allow for an extensive temporal, soil water monitoring network to be in place.

Although similar results were obtained with a spatially structured simulated error surface, it appeared that our optimized selection could not always be ranked low in terms of the MSE when compared with the most suitable random selection. This is most obvious for field HU, where the MSE for the optimum selection was close to the median value of the MSE's for random selections. This suggests once again that the spatial spread of monitoring sites across the field can avoid bias in regression analysis due to monitoring only for overestimated, or underestimated, parts of the field.

In addition, it should be noted that when the number of parameters was lower than four (reduced model), the maximum MSE value was lower as compared to the full model. This means that a small number of parameters could reduce the overall error when these parameters have a high level of uncertainties. From a practical point of

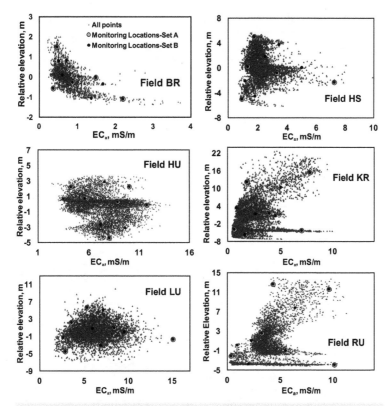

Figure 6. Maps relationship between ECa and field elevation for six fields.

Table 3. Summary of MSE estimates.

		BR	HS	HU	KR	LU	RU
ES 1 and 5 monitoring locations							
Random Selection	Min	0.05	0.04	0.04	0.04	0.05	0.05
	Median	0.59	0.19	0.15	0.41	0.21	0.61
Optimized Selection		0.07	0.06	0.10	0.06	0.07	0.06
ES 1 and 10 monitoring locations							
Random Selection	Min	0.04	0.04	0.04	0.04	0.04	0.04
	Median	0.08	0.07	0.07	0.08	0.06	0.07
Optimized Selection		0.07	0.05	0.06	0.05	0.06	0.06
ES 2 and 5 monitoring locations							
Random Selection	Min	0.04	0.05	0.04	0.04	0.04	0.04
	Median	0.48	0.22	0.18	0.34	0.16	0.28
Optimized Selection		0.08	0.05	0.19	0.05	0.08	0.04
ES 2 and 10 monitoring locations							
Random Selection	Min	0.04	0.04	0.04	0.04	0.04	0.03
	Median	0.07	0.06	0.06	0.07	0.06	0.07
Optimized Selection		0.04	0.05	0.05	0.05	0.04	0.05
ES 3 and 5 monitoring locations							
Random Selection	Min	0.05	0.04	0.04	0.04	0.05	0.05
	Median	0.36	0.17	0.16	0.39	0.22	0.33
Optimized Selection		0.09	0.07	0.06	0.05	0.14	0.05
ES 3 and 10 monitoring locations							
Random Selection	Min	0.04	0.04	0.04	0.04	0.04	0.04
	Median	0.06	0.07	0.07	0.08	0.07	0.07
Optimized Selection		0.06	0.05	0.05	0.05	0.05	0.04

Table 4. Percentage of random selections with higher MSE as compared to the optimized selection.

	BR	HS	HU	KR	LU	RU
Five (Set A) soil water monitoring sites						
ES 1	91	94	87	91	100	94
ES 2	86	100	48	95	80	95
ES 3	80	77	91	98	65	100
Ten (Set A & B) soil water monitoring sites						
ES 1	75	64	63	82	83	48
ES 2	90	84	91	75	85	69
ES 3	51	89	76	88	85	93

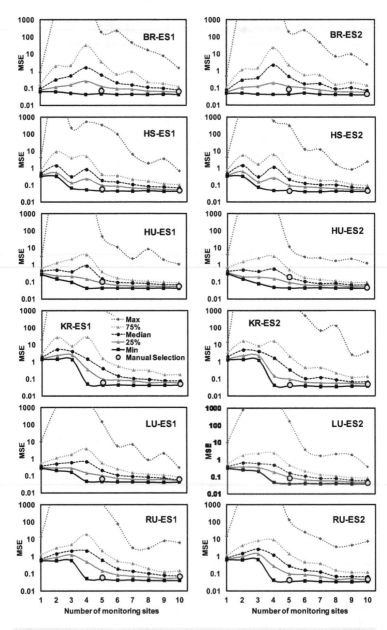

Figure 7. Summary of MSE values for random selection and optimized selection methods for all six fields using simulated surfaces based on error surface (ES 1 and ES 2).

view, this conclusion means that even a small number of monitoring sites can result in a quantification of soil water stress of relatively good quality in every part of the field, as long as these sites are selected carefully and water stress response to soil texture and/or landscape position is well defined. On the contrary, using a small number of monitoring sites to define parameters of a complex regression model greatly increases the chance of errors in estimating these parameters which would lead to lower WSP predictability.

Depending on the cost of added monitoring sites and the profit-reducing effect of WSP prediction errors, the optimum number of these sites will be different. Thus, errors in estimating water stress levels will lead to yield loss due to water stress and/or the extra cost of unnecessary water supply. Based on Equation (5), we anticipate that a single monitoring site placed in the most representative part of the field is a valid starting point. When soil variability, or field topography, presents a case for a substantial differences in expected plant available water storage, two monitoring locations are required. Once both soil EC_a and field elevation change substantially and do not correlate, three monitoring locations are needed. Once the interaction between the two high-density data layers is also influential, this supports the addition of a fourth monitoring site. However, due to the fact that the relationship between high-resolution data and WSP (Equation (5)) is not certain, extra monitoring sites could assure a more accurate definition of regression parameters, which should be fewer in number than the number of monitoring locations.

Based on this research, the greater the number of monitoring locations that can be equipped, the lower probability of selecting a set of locations with poor quality of regression parameters prediction will be. Once the number of monitoring locations is large, it is not as critical to seek the most appropriate locations as compared to a low number of such sites. With a relatively small number of temporal soil water monitoring sites (due to economic restrictions), the optimized site selection strategy described above can be used as a starting point when developing practical algorithms for irrigators pursuing the benefits of site-specific water management. However, the actual network design algorithm must involve the node communication capabilities, landscape topography, irrigation system geometry, and other infrastructure-based constraints. Some of these network architecture constraints are discussed in [14].

4. Conclusion

In this study, apparent soil electrical conductivity and field elevation data layers were mapped using on-the-go soil sensing technology. Both data layers were associated with soil water holding capacity. To assess the effect of monitoring site selection, various model WSP surfaces were obtained by adding up the calculated WSP and an error simulated according to different spatial distribution patterns. The random selection method and the optimized selection method based on specified criteria were applied and compared on the basis of the MSE computed from the predicted versus simulated WSP surfaces. The optimized selection of monitoring sites was helpful in reducing the chance of selecting monitoring sites with a poor capacity to recover the WSP regression model. Our results highlight the importance of covering the entire range of spatial data indicating water storage with a high resolution and spreading the sites across the field to account for any additional factors affecting water storage predictability.

Acknowledgements

This research was supported in part by funds provided through the Nebraska Center for Energy Science Research through the Water, Energy and Agriculture Initiative (WEAI); National Science and Engineering Research of Canada (NSERC) Discovery Grant; Agricultural Research Division of the University of Nebraska-Lincoln; and Deere and Company.

References

[1]   Viscarra Rossel, R.A., Adamchuk, V.I., Sudduth, K.A., McKenzie, N.J. and Lobsey, C. (2011) Proximal Soil Sensing: An Effective Approach for Soil Measurements in Space and Time, Chapter 5. *Advances in Agronomy* **113**, 237-283. http://dx.doi.org/10.1016/B978-0-12-386473-4.00005-1

[2]   Hedley, C. and Yule, I. (2009) A Method for Spatial Prediction of Daily Soil Water Status for Precise Irrigation Scheduling. *Agricultural Water Management*, **96**, 1737-1745. http://dx.doi.org/10.1016/j.agwat.2009.07.009

[3]   Kim, Y., Evans, R.G. and Iversen, W. M. (2009) Evaluation of Closed-Loop Site-Specific Irrigation with Wireless Sensor Network. *Journal of Irrigation and Drainage Engineering*, **135**, 25-31. http://dx.doi.org/10.1061/(ASCE)0733-9437(2009)135:1(25)

[4] Omary, M., Camp, C.R. and Sadler, E.J. (1997) Center Pivot Irrigation System Modification to Provide Variable Water Application Depths. *Applied Engineering in Agriculture*, **13**, 235-239. http://dx.doi.org/10.13031/2013.21604

[5] Miranda, F.R., Yoder, R. and Wilkerson, J.B. (2003) A Site-Specific Irrigation Control System. ASABE, St. Joseph, Paper No. 031129.

[6] Han, Y.J., Khalilian, A., Owino, T.O., Farahani, H.J. and Moore, S. (2009) Development of Clemson Variable-Rate Lateral Irrigation System. *Computers and Electronics in Agriculture*, **68**, 108-113. http://dx.doi.org/10.1016/j.compag.2009.05.002

[7] Lesch, S.M. (2005) Sensor-Directed Spatial Response Surface Sampling Designs for Characterizing Spatial Variation in Soil Properties. *Computers and Electronics in Agriculture*, **46**, 153-180. http://dx.doi.org/10.1016/j.compag.2004.11.004

[8] Minasny, B and McBratney, A.B. (2006) A Conditioned Latin Hypercube Method for Sampling in the Presence of Ancillary Information. *Computers and Geosciences*, **32**, 1378-1388. http://dx.doi.org/10.1016/j.cageo.2005.12.009

[9] Brus, D.J. and Heuvelink, G.B.M. (2007) Optimization of Sample Patterns for Universal Kriging of Environmental Variables. *Geoderma*, **138**, 86-95. http://dx.doi.org/10.1016/j.geoderma.2006.10.016

[10] de Gruijter, J.J., McBratney, A.B. and Taylor, J. (2010) Sampling for High-Resolution Soil Mapping. Proximal Soil Sensing. Springer Netherlands, 3-14.

[11] Adamchuk, V.I., Viscarra Rossel, R.A., Marx, D.B. and Samal, A.K. (2011) Using Targeted Sampling to Process Multivariate Soil Sensing Data. *Geoderma*, **163**, 63-73. http://dx.doi.org/10.1016/j.geoderma.2011.04.004

[12] Pan, L., Adamchuk, V.I., Martin, D.L., Schroeder, M.A. and Ferguson, R.B. (2013) Analysis of Soil Water Availability by Integrating Spatial and Temporal Sensor-Based Data. *Precision Agriculture*, **14**, 414-433. http://dx.doi.org/10.1007/s11119-013-9305-x

[13] SAS (2008) Optimality Criteria. SAS/QC User's Guide, SAS Institute Inc., Cary.

[14] An, W., Ci, S., Luo, H., Wu, D., Adamchuk, V., Sharif, H., Wang, X. and Tang, H. (2013) Effective Sensor Deployment Based on Field Information Coverage in Precision Agriculture. *Wireless Communications and Mobile Computing*. http://dx.doi.org/10.1002/wcm.2448

11

A Stakeholder-Guided Collaborative Approach to Improve Water Quality in a Nutrient Surplus Watershed

Hector German Rodriguez[1*], Jennie Popp[1], Edward Gbur[2], John Pennington[3]

[1]Department of Agricultural Economics and Agribusiness, University of Arkansas, Fayetteville, USA
[2]Agricultural Statistics Laboratory, University of Arkansas, Fayetteville, USA
[3]Beaver Watershed Alliance, Fayetteville, USA
Email: [*]hrodrig@uark.edu, jhpopp@uark.edu, egbur@uark.edu, john@beaverwatershedalliance.org

Abstract

The need for water quality improvement in nutrient surplus watersheds is a pressing issue on the agenda of some government agencies and environmental organizations. Including the water quality perceptions of different affected stakeholder groups in the decision-making process may help in addressing this issue. Unfortunately, there is a lack of published research focusing specifically on understanding how Arkansas stakeholders' perceptions of water quality issues can be used to build and implement comprehensive and workable water quality management plans. Therefore, the objective of this study was to use a stakeholder-guided collaborative approach to help research and outreach personnel to understand water quality perceptions of key stakeholders and to integrate stakeholder engagement in both the decision-making process and in the implementation of water quality management strategies within the Lincoln Lake Watershed in northwest Arkansas. Two key stakeholder groups (*i.e.*, Locals—residents and agricultural producers—and Outsiders—water quality specialist across the state) were surveyed to assess their perceptions regarding: 1) causes of watershed water quality problems, 2) parties responsible for water quality improvement, 3) effectiveness and affordability of best management practices to reduce water quality degradation, and 4) the stakeholders' interactions with county, state and federal government. A total of 209 complete surveys (49% response rate) were received. Survey responses were compared to determine if significant differences existed between the two stakeholder groups' perceptions of water quality performing Fisher's exact tests. Results from the study showed that water quality is still perceived as an issue in the Lincoln Lake Watershed. Significant differences were found between the two stakeholder groups' perceptions regarding: 1) different groups' contributions to water degradation, 2) groups' responsibilities for cleanup, 3) effectiveness of five best management practices, 4) affordability of four best management practices, and 5) what level

---

[*]Corresponding author.

of government (*i.e.*, county, state, federal) best represents Locals' water quality needs and concerns. The lessons learned from this collaborative approach helped identifying Locals' important knowledge gaps regarding water quality and best management practices effectiveness. Consequently, awareness and education campaigns in conjunction with a stewardship recognition program were conducted to encourage appropriate water conservation strategies within the Lincoln Lake watershed and its adjacent areas.

Keywords

Stakeholders' Perceptions, Environmental Decision Making, Best Management Practices, Nutrient Surplus Watershed, Watershed Management

1. Introduction

1.1. Water Quality Improvement and the Decision-Making Process

Water quality improvements facilitated by policy action is pressing need in many nutrient surplus watersheds. In recent years, political and technical factors (*i.e.*, water quality modeling/monitoring of nutrient, pesticide and sediment concentrations in streams and rivers) have influenced environmental policy changes in the United States [1]-[5]. Frameworks for effective environmental policy are complex because they must deal simultaneously with environmental, economic, and social concerns and the varying interests of different stakeholders [3] [6].

One further complicating factor is the difference between the perception and reality of stakeholders' inclusion in the decision-making process. The inclusion of several stakeholder groups, with their different interests and perceptions, contributes to improve the decision-making process [7]. However, due to the complexities mentioned above some stakeholders may be excluded [8]. Examples of this phenomenon can be seen when reviewing water quality management systems [3].

Policy goals, institutional goals and interest group goals may be viewed differently depending on the perspectives and biases of those involved in the formulation and implementation processes [9]. Additionally, stakeholders' interests vary based on, among other things, the potential favorable or unfavorable outcome obtained from a policy change and its direct and indirect effects. In fact, key stakeholders could oppose a water policy that has potential to adversely affect them [6].

Although regulation is a common policy tool used to address water pollution issues [2] [10]-[13], several government agencies encourage the implementation of best management practices (BMPs) through a voluntary adoption approach [2] [4] [11] [13]. From the institutional point of view, agricultural producers, other watershed residents and the public could benefit greatly and directly from a behavioral change toward better participation in the water conservation programs provided by county, state or federal agencies [1] [2] [14]. However, collective action depends on trust (Imperial 2005). So, if some in the group cooperate, others are more likely to cooperate too [2]. For instance, by providing cost-share programs and delivering information about the benefits of BMP adoption, county, state and federal organizations could improve trust and cooperation from stakeholders [11] [13].

1.2. Water Quality Management in Arkansas

The perception of water problems can differ across stakeholder groups [2] [6] [15]. Even with consensus that a water quality problem exists, each group can have its own objectives for management that may conflict with the objectives of one or more other groups [16]. In nutrient surplus watersheds in Arkansas, there are numerous groups with an interest in addressing water quality issues. In fact, there are more than twenty organizations with responsibility for preserving the state's water quantity, quality and public health [17]. Despite considerable effort to include key stakeholders in the decision-making process of preserving the quality of water resources in Arkansas, there is a lack of published research focusing specifically on understanding how Arkansas stakeholders' perceptions of water quality issues can be used to build and implement comprehensive and workable water quality management plans.

In 2006, the University of Arkansas (UA) Division of Agriculture was awarded a USDA CSREES Conservation Effects Assessment Project (CEAP) grant. The overall purpose of this multidisciplinary project was to understand the relationship between BMPs and water quality in the Lincoln Lake watershed (LLW) in Northwest Arkansas. Hoag *et al.* gives a more detailed description of this project [18]. A specific goal was to assess the perceptions of different LLW stakeholders as well as county, state, and federal water quality specialists and regulators on water quality and the effectiveness of agricultural BMPs. This was a multi-institution effort that also included Purdue University, the Natural Resource Conservation Service, the Arkansas Natural Resource Commission, and Association of Conservation Districts.

1.3. Research Purpose

This study presents the results of a stakeholder-guided collaborative approach used to help research and outreach personnel to understand water quality perceptions of key stakeholders and to integrate stakeholder engagement in both the decision-making process and in the implementation of water quality management strategies. Two key stakeholder groups were surveyed to assess their perceptions regarding: 1) causes of watershed water quality problems, 2) parties responsible for water quality improvement, 3) effectiveness and affordability of BMPs to reduce water quality degradation, and 4) the stakeholders' interactions with county, state and federal government in setting and enforcing relevant water quality management policies. The lessons learned from this collaborative approach helped to identify appropriate strategies to increase awareness and implementation of BMPs enhancing the problem-solving capacity of stakeholders in this troubled watershed.

2. Materials and Methods

The Lincoln Lake Watershed (LLW)

The LLW in Northwest Arkansas is a subbasin of the Illinois River watershed that includes areas of both Northwest Arkansas and Northeast Oklahoma. The drainage area of the LLW is approximately 32 km^2 (12 mi^2). The dominant agricultural activities in the watershed are beef cattle and poultry operations. For a more detailed description of this watershed, see [19]. Nonpoint source (NPS) transport of nutrients (especially phosphorous), sediment, and pathogens from surface applied animal manure is a major concern [20] [21].

For this study, key stakeholders were divided into two groups: 1) watershed Locals and 2) Outsiders. Locals were comprised of two subgroups: agricultural producers (who own or operate land within the watershed) and other watershed residents (defined as households, business owners and other landowners). Outsiders were defined as natural resource specialists or water regulators from several county, state and federal institutions within the state who have knowledge of, or authority for, water quality management in the watershed.

Washington County Tax Assessor's Office records were used to identify 75 agricultural producers and 243 residents (Locals) in the LLW. One hundred and sixty water quality specialists (Outsiders) were identified through appropriate government agencies and academic institutions within the state. A stakeholder focus group, consisting of a county judge, five farmers, and two residents from an adjoining subwatershed, was formed to advise the project team on data collection and analysis. Survey data were collected in three different ways.

First, agricultural producers were interviewed in person by UA Cooperative Extension Service personnel. Second, residents were asked to complete a survey during watershed stakeholder meetings. Surveys were later mailed to residents absent from these meetings. Finally, specialists were mailed their surveys.

The survey questionnaires consisted of four or six sections depending on the stakeholder group surveyed. Some questions were developed specifically to gather the perceptions and behaviors of each stakeholder group regarding the four primary issues listed previously. However, when it was pertinent, the information was disaggregated to compare the Locals' subgroups (*i.e.*, compare agricultural producers to other watershed residents) or to compare a Locals' subgroup to the Outsiders (*i.e.*, compare agricultural producers to water quality specialists). Survey responses were compared to determine if significant differences existed between the two stakeholder groups' perceptions of water quality performing Fisher's exact tests to compare the distributions of responses from the two stakeholder groups.

3. Results

3.1. Survey Demographic Analysis

In total, of the 478 surveys distributed either by mail or in person, 209 usable responses (49%) were received.

From the Locals group, 131 survey responses were received. Sixty-three out of the 75 agricultural producers (84%) completed the survey. Their primary agricultural activities included beef cattle, hay, pasture, and broiler production. Sixty-eight out of 243 residents (28%) completed the survey. The majority of these respondents (81%) were individuals who had their primary residence in the watershed.

Seventy-eight out of 160 (or 49%) of the Outsiders completed the questionnaire. They represented six institutions: the UA Division of Agriculture Experiment Station (11%), UA Division of Agriculture Cooperative Extension Service (29%), Arkansas Conservation Districts (8%), Natural Resources Conservation Service (25%), Arkansas Department of Environmental Quality (8%) and the Arkansas Natural Resources Commission (19%).

3.2. Perceptions Specifically Regarding Water Quality Problems in the Lincoln Lake Watershed

Several studies have shown long-term trend improvements in water quality in the watershed since mid-1990s [22] and early 2000s [23]. Two recent studies have shown that overall flow-adjusted concentrations of phosphorus (the main pollutant of concern in LLW) have been decreasing from the Beatty Branch, Upper Moores Creek, and Moores Creek from 1992 to 2007 [19] and the Illinois River watershed from 1997 to 2010 [24]. However, the survey findings showed that the perceptions of water quality in the LLW may differ from this reality.

Locals and only those Outsiders familiar with the LLW (*i.e.*, Lincoln Lake, Moores Creek and Beatty Branch) were asked their perceptions regarding the existence of water quality problems in the watershed. Of the 131 Locals respondents, 129 individuals answered this question. Only 18 out of 78 Outsiders (23%) were familiar enough with the LLW to answer the survey questions specifically related to the LLW. A Fisher's exact test found significant differences between the distributions of responses of Locals and Outsiders regarding their perception of the water quality in each of the three bodies of water analyzed: Lincoln Lake, Beatty Branch and Moores Creek. Approximately 26% of the Locals and 72% of the Outsiders agreed that water quality problems existed in Lincoln Lake (**Table 1**).

Similarly for all three bodies of water, a greater percentage of Outsiders perceived that water quality problems existed than Locals did. Nevertheless, when groups were asked about their perceptions regarding different water uses (*i.e.*, drinking, fishing, and swimming) in the three water bodies, no significant differences between stakeholders' responses were found (**Tables 2-4**).

In fact, the majority of the Locals and Outsiders respondents believed the water quality in the LLW was acceptable for drinking, once treated (**Table 2**) and fishing (**Table 3**). Respondents seemed more neutral in their perception that water was good for swimming (**Table 4**).

3.3. Perceptions Regarding Water Degradation Contributions and Clean up Responsibilities

Locals and Outsiders were asked questions about contributions to, and responsibilities for, cleaning up water quality issues. Both groups were asked specifically about their perceptions of the contributions of different

Table 1. General perceptions that water quality problems exist in the Lincoln Lake watershed (percentage of respondents).

Level[a]	Lincoln Lake (p = 0.0010)[b]		Moores Creek (p = 0.0004)[c]		Beatty Branch (p = 0.0007)[d]	
	Locals	Outsiders	Locals	Outsiders	Locals	Outsiders
Agree	26.4	72.2	23.0	66.7	17.0	55.6
Neutral	36.4	16.7	39.3	27.8	45.5	38.9
Disagree	37.2	11.1	37.7	5.6	37.5	5.6

[a]Table entry as a percentage of respondents in that group selecting that response. [b]Response group includes 129 of the 131 Locals and 18 of 18 Outsiders. [c]Response group includes 122 of the 131 Locals and 18 of 18 Outsiders. [d]Response group includes 112 of the 131 Locals and 18 of 18 Outsiders.

Table 2. General perceptions that water is good for drinking (if treated) in the Lincoln Lake watershed, lakes and streams (percentage of respondents).

Level[a]	Lincoln Lake (p > 0.9999)[b]		Moores Creek (p = 0.2840)[c]		Beatty Branch (p = 0.5037)[d]	
	Locals	Outsiders	Locals	Outsiders	Locals	Outsiders
Agree	70.1	72.2	67.2	61.1	64.0	55.6
Neutral	19.7	16.7	23.8	16.7	27.2	27.8
Disagree	10.2	11.1	9.0	22.2	8.8	16.7

[a]Table entry as a percentage of respondents in that group selecting that response. [b]Response group includes 127 of the 131 Locals and 18 of 18 Outsiders. [c]Response group includes 122 of the 131 Locals and 18 of 18 Outsiders. [d]Response group includes 114 of the 131 Locals and 18 of 18 Outsiders.

Table 3. General perceptions that water is good for fishing in the Lincoln Lake watershed, lakes and streams (percentage of respondents).

Level[a]	Lincoln Lake (p = 0.1575)[b]		Moores Creek (p = 0.9328)[c]		Beatty Branch (p = 0.8154)[d]	
	Locals	Outsiders	Locals	Outsiders	Locals	Outsiders
Agree	75.4	57.9	52.5	50.0	47.0	44.4
Neutral	21.4	42.1	38.5	38.9	41.7	50.0
Disagree	3.2	0.0	9.0	11.1	11.3	5.6

[a]Table entry as a percentage of respondents in that group selecting that response. [b]Response group includes 126 of the 131 Locals and 18 of 18 Outsiders. [c]Response group includes 122 of the 131 Locals and 18 of 18 Outsiders. [d]Response group includes 115 of the 131 Locals and 18 of 18 Outsiders.

Table 4. General perceptions that water is good for swimming in the Lincoln Lake watershed, lakes and streams (percentage of respondents).

Level[a]	Lincoln Lake (p = 0.2796)[b]		Moores Creek (p = 0.7289)[c]		Beatty Branch (p = 0.7243)[d]	
	Locals	Outsiders	Locals	Outsiders	Locals	Outsiders
Agree	38.9	22.2	32.2	22.2	30.4	22.2
Neutral	36.5	55.6	42.2	50.0	44.4	55.6
Disagree	24.6	22.2	25.6	27.8	25.2	22.2

[a]Table entry as a percentage of respondents in that group selecting that response. [b]Response group includes 126 of the 131 Locals and 18 of 18 Outsiders. [c]Response group includes 121 of the 131 Locals and 18 of 18 Outsiders. [d]Response group includes 115 of the 131 Locals and 18 of 18 Outsiders.

groups to water quality problems in the LLW (**Table 5**). Again, only answers from Outsiders familiar with the LLW were used in this analysis.

Significant differences existed in stakeholder perceptions regarding different groups' contributions to water quality degradation. For instance, most Locals believed that agriculture either did not contribute at all (13%) or contributed only a small amount (64%) to water quality problems in the LLW. On the other hand, 61% of Outsiders believed that agriculture's contribution was large. Additionally, with the exception of new construction, less than 30% of the Locals believed any one source was largely responsible for water quality degradation while over 50% of the Outsiders felt four sources were largely responsible for water quality degradation.

Some similarities in perceptions can be drawn from the results. First, overall, both stakeholder groups believed that all listed sources in **Table 5** contributed at least small amounts to water degradation; the relative ranking of those sources as contributors between stakeholder groups was what differed (**Table 5**). New construction was chosen most often by both Locals and Outsiders as a large contributor of water quality degradation but the magnitude of agreement between Locals and Outsiders differed considerably (46% vs. 77% respectively).

As expected, while significant differences existed between Locals and Outsiders perceptions of which potential source group(s) was/were responsible for water pollution cleanup, some similarities were found too (**Table 6**). Both Locals and Outsiders believed that all potential source groups shared some responsibility for cleaning up. Locals and Outsiders both identified most often new construction, industry, and the city sewer system as holding large responsibilities for cleaning up, but the magnitude attributed to each potential source differed across stakeholder groups. However, nearly three times as many Outsiders (67%) than Locals (24%) believed that agricultural producers are largely responsible for cleanup.

Interesting changes were found in distribution of none, small and large between potential source and responsibility perceptions for both Locals and Outsiders. When comparing **Table 5** and **Table 6**, for any given poten-

Table 5. Perceptions of the amount of contribution of different groups to water quality problems in nutrient surplus areas including the Lincoln Lake watershed (percentage of respondents).

Group[a]	Locals			Outsiders			p-value
	None	Small	Large	None	Small	Large	
Agricultural Producers[b]	13.0	64.2	22.8	0.0	38.9	61.1	0.0036
City Sewer System[c]	16.2	55.6	28.2	0.0	33.3	66.7	0.0041
Households[d]	19.0	57.0	24.0	0.0	72.2	27.8	0.1074
Industry[e]	22.1	48.4	29.5	0.0	41.2	58.8	0.0142
New Construction[f]	14.9	39.7	45.5	0.0	23.5	76.5	0.0435
Outdoor Recreation[g]	52.5	40.7	6.8	16.7	83.3	0.0	0.0033

[a]Table entry as a percentage of respondents in that group selecting that response. [b]Response group includes 123 of the 131 Locals and 18 of 18 Outsiders. [c]Response group includes 117 of the 131 Locals and 18 of 18 Outsiders. [d]Response group includes 121 of the 131 Locals and 18 of 18 Outsiders. [e]Response group includes 122 of the 131 Locals and 18 of 18 Outsiders. [f]Response group includes 121 of the 131 Locals and 17 of 18 Outsiders. [g]Response group includes 118 of the 131 Locals and 18 of 18 Outsiders.

Table 6. Perceptions of responsibility of cleaning up of different groups to improve water quality in nutrient surplus areas including the Lincoln Lake watershed (percentage of respondents).

Group[a]	Locals			Outsiders			p-value
	None	Small	Large	None	Small	Large	
Agricultural Producers[b]	20.8	55.0	24.2	0.0	33.3	66.7	0.0007
City Sewer System[c]	18.1	49.1	32.8	0.0	22.2	77.8	0.0010
Households[d]	19.8	55.2	25.0	5.9	64.7	29.4	0.4582
Industry[e]	13.0	47.8	39.1	0.0	38.9	61.1	0.1218
New Construction[f]	11.9	39.0	49.2	0.0	22.2	77.8	0.0681
Outdoor Recreation[g]	47.3	42.0	10.7	16.7	83.3	0.0	0.0047

[a]Table entry as a percentage of respondents in that group selecting that response. [b]Response group includes 120 of the 131 Locals and 18 of 18 Outsiders. [c]Response group includes 116 of the 131 Locals and 18 of 18 Outsiders. [d]Response group includes 116 of the 131 Locals and 18 of 18 Outsiders. [e]Response group includes 115 of the 131 Locals and 18 of 18 Outsiders. [f]Response group includes 118 of the 131 Locals and 18 of 18 Outsiders. [g]Response group includes 112 of the 131 Locals and 18 of 18 Outsiders.

tial source except recreation, a greater percentage of both Locals and Outsiders respondents put more responsibility on these sources for cleanup than they did for water quality pollution. Similarly a greater percentage of both Locals and Outsiders said these sources were not responsible for cleanup compared to the percentages that believed these sources were not responsible for the water quality problems. It suggests that those who place small responsibility on each of these potential sources as causes of the degradation of water quality were split in their opinion as to how much responsibility to assign them for cleanup.

3.4. Perceptions Regarding Effectiveness and Affordability of Best Management Practices

Only Locals and those Outsiders familiar with the LLW were asked whether they agreed that each of 15 BMPs was effective in reducing sediment and/or nutrient loss from agricultural land. The possible responses were "yes", "no" and "not sure." The percentage responses of each group are presented in **Table 7**.

There are four important points to note about this analysis. First, for any of the 15 practices listed, 17 to 26 percent of the 149 potential respondents did not answer the question. Second, significant differences were found in Locals and Outsiders perceptions for 11 of the 15 BMPs. In general, a larger percentage of Outsiders thought the BMPs were effective while a larger percentage of Locals were unsure of BMP effectiveness. The reason for these differences may be because Outsiders are more likely to be exposed to BMPs through their work than Locals not directly engaged in agriculture.

Third, there were no significant differences between opinions of Locals and Outsiders for four of the 15 BMPs: controlled grazing, prescribed grazing, soil testing, and use of legumes to reduce N application. In general, both Locals and Outsiders believed that controlled grazing and soil testing were effective BMPs. But both groups were more uncertain about the effectiveness of prescribed grazing and using legumes to reduce N application. Fourth, the affordability of BMPs was directly asked to agricultural producers (a subset of the Locals) and only to Outsiders familiar with the LLW. In general, agricultural producers and Outsiders believed that agricultural producers can afford to adopt BMPs.

Table 7. Perceptions that best management practices are effective in reducing sediment and/or nutrient loss from agricultural lands (percentage of respondents).

Best Management Practice	Total[a] Responses	Locals			Outsiders			p-value
		No	Yes	Not Sure	No	Yes	Not Sure	
Controlled Grazing	121	10.7	75.7	13.6	0.0	94.4	5.6	0.2395
Filter Strips/Riparian Buffers	118	5.0	59.0	36.0	0.0	100.0	0.0	0.0014
Prescribed Grazing	117	16.2	53.5	30.3	0.0	72.2	27.8	0.1501
Pasture Grass Management	120	7.8	70.6	21.6	0.0	100.0	0.0	0.0264
Stream Bank Stabilization	117	13.1	47.5	39.4	5.6	94.4	0.0	0.0003
Cattle Track Stabilization	110	10.8	47.3	41.9	0.0	88.2	11.8	0.0066
Stream Fencing	116	18.4	49.0	32.6	0.0	88.9	11.1	0.0054
Basin Fertilizer Application	123	5.7	73.3	21.0	0.0	100.0	0.0	0.0376
Litter Storage Shed	119	16.8	49.5	33.7	22.2	77.8	0.0	0.0039
Manure Composting	114	9.4	49.0	41.6	27.8	55.6	16.6	0.0356
Soil Testing	124	5.7	80.2	14.2	11.1	88.9	0.0	0.1482
Use of Legumes to Reduce N Application	114	6.2	60.8	33.0	23.5	58.8	17.7	0.0595
Nutrient Management Plan	118	9.0	52.0	39.0	5.6	88.9	5.5	0.0058
Manure instead of Commercial Fertilizer	121	7.8	68.0	24.3	27.8	72.2	0.0	0.0037
Waste Treatment Lagoon	116	19.4	17.4	63.3	11.1	77.8	11.1	<0.0001

[a]Table entry as a percentage of respondents in that group selecting that response.

Significant differences in perceptions regarding only two BMPs (nutrient management plan (NMP) and waste treatment lagoon) were found (**Table 8**). Surprisingly, only 68% of the agricultural producers believed that NMPs are affordable. This is a low cost practice encouraged by county, state, and federal agencies and institutions. It could have been possible that the agricultural producers and perhaps, some producers thought that the NMP was unaffordable because it could have limited how much they fertilized, which could have reduced their yield, and as a result their carrying capacity or hay sales. According to J. Pennington (personal communication, January 30, 2014), even though conservation plans have been around for decades, the transition from the conservation plan to the NMP was still new among some of the population and they could have been unfamiliar with the new process.

3.5. Perceptions Regarding Interactions with the Government

For this section of the survey, the Locals group was disaggregated into agricultural producers (**Table 9**) and residents (**Table 10**) to ask each group specific perceptions regarding what level of government (*i.e.*, county, state, federal) best represents their water quality needs and concerns.

Outsiders were also asked their perceptions of what level of government they thought represented agricultural producers' and residents' water quality needs and concerns best. A Fisher's exact test found a significant difference between agricultural producers' and Outsiders' responses ($p < 0.0001$). Agricultural producers indicated that the county government represented them best while Outsiders believed that all levels—county, state and federal—of government represented agricultural producers' interests (**Table 9**) about equally well.

The reasons for these differences could be that while agricultural producers generally have more direct interaction with officials on the county level, the Outsiders surveyed represented all levels of government and are working to advance agriculture in the state as a whole. On the other hand, no significant difference in opinions between residents and Outsiders was found ($p = 0.5252$). Both residents and Outsiders believed that the county government represented residents' water quality interests and concerns best (**Table 10**).

Table 8. Perceptions that agricultural producers can afford adopting best management practices in their lands (percentage of respondents).

Group	Total[a] Responses	Agricultural Producers		Outsiders		p-value
		No	Yes	No	Yes	
Controlled Grazing	71	24.5	75.5	5.6	94.4	0.0981
Filter Strips/Riparian Buffers	68	42.0	58.0	38.9	61.1	>0.9999
Prescribed Grazing	59	27.9	72.1	6.3	93.7	0.0903
Pasture Grass Management	71	18.9	81.1	16.7	83.3	>0.9999
Stream Bank Stabilization	73	76.4	23.6	83.3	16.7	0.7453
Cattle Track Stabilization	65	38.0	62.0	53.3	46.7	0.3739
Stream Fencing	73	78.2	21.8	61.1	38.9	0.1316
Basin Fertilizer Application	70	17.3	82.7	11.1	88.9	0.7162
Litter Storage Shed	72	64.8	35.2	55.6	44.4	0.5770
Manure Composting	61	47.7	52.3	41.2	58.9	0.7767
Soil Testing	74	8.9	91.1	0.0	100.0	0.3263
Use of Legumes to Reduce N Application	65	17.0	83.0	38.9	61.1	0.0977
Nutrient Management Plan	68	32.0	68.0	5.6	94.4	0.0288
Manure instead of Commercial Fertilizer	69	7.8	92.2	11.1	88.9	0.6473
Waste Treatment Lagoon	58	85.7	14.3	56.2	43.8	0.0310

[a]Table entry as a percentage of respondents in that group selecting that response.

Table 9. Perceptions that a specific level of government represents agricultural producers water quality needs and concerns best (percentage of respondents).

Government Level[a]	Agricultural Producers	Outsiders
County	83.1	36.2
State	13.6	37.7
Federal	3.4	26.1

[a]Table entry as a percentage of respondents in that group selecting that response. Response group includes 59 of the 63 agricultural producers and 69 of 78 Outsiders.

Table 10. Perceptions that a specific level of government represents the residents' water quality needs and concerns best (percentage of respondents).

Government Level[a]	Residents	Outsiders
County	69.2	60.3
State	23.1	26.5
Federal	7.7	13.2

[a]Table entry as a percentage of respondents in that group selecting that response. Response group includes 52 of the 68 residents and 68 of 78 Outsiders.

Agricultural producers and Outsiders were asked their opinion of whether additional power/authority should be given to each level of government to address water quality in the LLW. While no significant differences were found, the data did provide the following insights. The greatest percentages of agricultural producers and Outsiders believed that County government should be given at least some additional power to address water quality concerns (**Table 11**).

The smallest percentages of agricultural producers and Outsiders believed Federal government should receive some or a lot of additional power. Interestingly as well, 71% of Outsiders vs. only 48% of agricultural producers thought some or a lot of additional power should be given to state government. This may be because state government offices oversee most of the water quality regulation and a large percentage of the surveyed Outsiders are employed by state government.

The perceived importance of the county government over the state government by the residents group could be explained by the fact that residents were more willing to give additional power to a government entity that is familiar with their local water quality problems and consequently, could represent their water quality needs and concerns better (**Table 12**).

Although no significant differences were found (at the 5% level of significance) when Residents and Outsiders were asked about giving additional power to the state government, 59% of the residents felt that the state government should not receive additional power. Only 29% of the outsiders felt in the same way. In fact, 24% of the Outsiders believed that the state government should receive a lot of additional power compared to 9% of the residents. However, both groups' opinions were very similar regarding the federal government.

4. Discussion

The survey analysis fulfilled the goal of providing research and outreach personnel with information regarding the perceptions of key stakeholders regarding water quality. In many cases, it revealed significant differences in perceptions between those who lived and worked in the watershed (Locals) and water quality specialists (Outsiders). Gaps in knowledge regarding water quality and BMPs were identified and addressed by the CEAP

Table 11. Perceptions that additional power/authority should be given to each level of government to address water quality (Agricultural Producers vs. Outsiders).

Government[a]	Agricultural Producers			Outsiders			p-value
	None	Some	A lot	None	Some	A lot	
County	26.6	31.7	41.7	41.2	23.5	35.3	0.6104
State	51.7	36.7	11.6	29.4	47.1	23.5	0.2116
Federal	81.7	13.3	5.0	70.1	23.5	5.9	0.5652

[a]Table entry as a percentage of respondents in that group selecting that response. Response group includes 60 of the 63 agricultural producers and 17 of 18 Outsiders.

Table 12. Perceptions that additional power/authority should be given to each level of government to address water quality (Residents vs. Outsiders).

Government[a]	Residents			Outsiders			p-value
	None	Some	A lot	None	Some	A lot	
County	32.8	31.0	36.2	41.2	23.5	35.3	0.7883
State	58.9	32.1	9.0	29.4	47.1	23.5	0.0661
Federal	69.2	28.9	1.9	70.1	23.5	5.9	0.6360

[a]Table entry as a percentage of respondents in that group selecting that response. Response group includes 55 of the 63 agricultural producers and 17 of 18 Outsiders.

project team through extensive outreach efforts in the LLW that included over two dozen community based meetings attended by more than 1000 diverse stakeholders. Through these meetings, science based information regarding: 1) causes of water quality problems, 2) responsibilities for water quality improvement, and 3) effectiveness and affordability of BMPs were shared.

One important survey result for the CEAP team was the perception by some agricultural producers that NMPs were not affordable. In fact, only 39% of the LLW agricultural producers had implemented a NMP at the time they were surveyed. Based on these survey results, the CEAP project team focused in part on educational meetings, face-to-face interviews and farm visits to promote a sign up campaign for NMP implementation. This campaign was a cooperative effort between the Natural Resources Conservation Services, the Washington County Conservation District, and the Washington County Cooperative Extension Service. The results of this campaign included over 1000 producers in Washington County and several surrounding counties who were exposed to information regarding NMPs and BMPs. Additionally, NMP adoption within the watershed increased by 66% (45 new NMPs) and 77 other NMPs were adopted adjacent to the LLW. Furthermore, 3525 ha (8710 ac), in and adjacent to the watershed were soil sampled.

A second important survey result also pointed to differences in perception between Locals and Outsiders regarding BMP effectiveness. The CEAP team conducted a BMP awareness and education campaign in conjunction with a stewardship recognition program. Through this campaign, 25 agricultural producers in the LLW implemented at least one new BMP on their farms. Additionally, 54 participants who were implementing BMPs on their farms received a stewardship recognition sign that was displayed by a road adjacent to their farm. The goal of these signs was two-fold: 1) to recognize producers who were implementing BMPs and 2) to increase awareness of BMP use on farms to others in the community.

A third important survey result indicated that Locals, especially agricultural producers, felt disengaged from water conservation policy. Specifically, agricultural producers perceived that county officials represented their needs better than state and federal officials did. Yet these county officials needed more power to address water quality concerns. The CEAP project team shared the results with watershed organizations, state and federal agencies, and other stakeholders across the state of Arkansas. Subsequent input from agricultural producers to local officials have resulted in changes to some county level conservation program efforts and increased producer participation. However, at the same time millions of dollars available through a non-county level conservation program in the watershed go unused; producers state that the program (which did not include producer input) does not meet their needs.

Thanks in large part due to information gleamed from the survey, the CEAP team was able to mount a successful education campaign within LLW that raised awareness of water quality issues and effected the adoption of BMPs. The impacts were not limited to LLW Locals stakeholders. Others recognizing the success in LLW adopted the CEAP model to provide watershed relevant water quality education and the enactment of the "stewardship sign program" in the greater Illinois River Watershed and War Eagle Watershed of the White River.

5. Conclusions

Analysis of the survey responses provided timely information regarding the water quality perceptions of key stakeholders in the LLW. By sharing survey results with Locals stakeholders and Outsiders, the CEAP project team improved stakeholders' inclusion, trust, cooperation and base knowledge about causes and potential solutions to water quality degradation.

This study provides a list of lessons learned:

- Availability of technical information regarding water quality, BMPs' effectiveness and BMP affordability does not guarantee that all types of stakeholders will be familiar with BMPs, even in a nutrient surplus watershed.
- Understanding stakeholders' water quality perceptions and using that information to create education and outreach programs is a first step in addressing the water quality issue.
- Engagement of key stakeholders is fundamental in the development of local policies to water quality problems.
- Educational programs, developed through assessments of stakeholder perceptions can results in actions that can improve water quality in a nutrient surplus watershed.

Lessons learned from this collaborative approach could be transferred to other nutrient surplus watersheds to develop education campaigns and stewardship recognition programs to improve water quality in the region.

Acknowledgements

Funding for this research was provided in part by the USDA-CSREES under Conservation Effects Assessment Project Grant Program (award 2005-04333).

References

[1]   Imperial, M.T. (2005) Using Collaboration as a Governance Strategy: Lessons from Six Watershed Management Programs. *Administration & Society*, **37**, 281-320. http://dx.doi.org/10.1177/0095399705276111

[2]   Lubell, M. and Fulton, A. (2008) Local Policy Networks and Agricultural Watershed Management. *Journal of Public Administration Research and Theory*, **18**, 673-696. http://dx.doi.org/10.1093/jopart/mum031

[3]   Osmond, D., Meals, D., Hoag, D., Arabi, M., Luloff, A., Jennings, G. and Line, D. (2012) Improving Conservation Practices Programming to Protect Water Quality in Agricultural Watersheds: Lessons Learned from the National Institute of Food and Agriculture—Conservation Effects Assessment Project. *Journal of Soil and Water Conservation*, **67**, 122A-127A. http://dx.doi.org/10.2489/jswc.67.5.122A

[4]   Savage, J. and Ribaudo, M.O. (2013) Impact of Environmental Policies on the Adoption of Manure Management Practices in the Chesapeake Bay Watershed. *Journal of Environmental Management*, **129**, 143-148. http://dx.doi.org/10.1016/j.jenvman.2013.06.039

[5]   Scheberle, D. (2004) Federalism and Environmental Policy: Trust and the Politics of Implementation. Georgetown University Press, Washington DC.

[6]   Grimble, R. and Wellard, K. (1997) Stakeholder Methodologies in Natural Resource Management: A Review of Principles, Contexts, Experiences and Opportunities. *Agricultural Systems*, **55**, 173-193. http://dx.doi.org/10.1016/S0308-521X(97)00006-1

[7]   Ravnborg, H.M. and Westermann, O. (2002) Understanding Interdependencies: Stakeholder Identification and Negotiation for Collective Natural Resource Management. *Agricultural Systems*, **73**, 41-56. http://dx.doi.org/10.1016/S0308-521X(01)00099-3

[8]   Mostashari, A. and Sussman, J. (2005) Stakeholder-Assisted Modeling and Policy Design Process for Environmental Decision-Making. *Journal of Environmental Assessment Policy & Management*, **7**, 355-386.

http://dx.doi.org/10.1142/S1464333205002110

[9]   Brugnach, M., Dewulf, A., Henriksen, H.J. and van der Keur, P. (2011) More Is Not Always Better: Coping with Ambiguity in Natural Resources Management. *Journal of Environmental Management*, **92**, 78-84. http://dx.doi.org/10.1016/j.jenvman.2010.08.029

[10] Barnes, A.P., Willock, J., Hall, C. and Toma, L. (2009) Farmer Perspectives and Practices Regarding Water Pollution Control Programmes in Scotland. *Agricultural Water Management*, **96**, 1715-1722. http://dx.doi.org/10.1016/j.agwat.2009.07.002

[11] Barnes, A.P., Willock, J., Toma, L. and Hall, C. (2011) Utilising a Farmer Typology to Understand Farmer Behaviour towards Water Quality Management: Nitrate Vulnerable Zones in Scotland. *Journal of Environmental Planning and Management*, **54**, 477-494. http://dx.doi.org/10.1080/09640568.2010.515880

[12] Borisova, T., Racevskis, L. and Kipp, J. (2012) Stakeholder Analysis of a Collaborative Watershed Management Process: A Florida Case Study. *Journal of the American Water Resources Association*, **48**, 277-296. http://dx.doi.org/10.1111/j.1752-1688.2011.00615.x

[13] Kay, P., Grayson, R., Phillips, M., Stanley, K., Dodsworth, A., Hanson, A., Walker, A., Foulger, M., McDonnell, I. and Taylor, S. (2012) The Effectiveness of Agricultural Stewardship for Improving Water Quality at the Catchment Scale: Experiences from an NVZ and ECSFDI Watershed. *Journal of Hydrology*, **422-423**, 10-16. http://dx.doi.org/10.1016/j.jhydrol.2011.12.005

[14] Williams, J.R., Smith, C.M., Roe, J.D., Leatherman, J.C. and Wilson, R.M. (2012) Engaging Watershed Stakeholders for Cost-Effective Environmental Management Planning with "Watershed Manager." *Journal of Natural Resources & Life Sciences Education*, **41**, 44-53.

[15] Popp, J., Rodríguez, H.G., Gbur, E. and Pennington, J. (2007) The Role of Stakeholders' Perceptions in Addressing Water Quality Disputes in an Embattled Watershed. *Journal of Environmental Monitoring and Restoration*, **3**, 255-263. http://dx.doi.org/10.4029/2007jemrest3no125

[16] Sang, N. (2008) Informing Common Pool Resource Problems: A Survey of Preference for Catchment Management Strategies amongst Farmers and the General Public in the Ythan River Catchment. *Journal of Environmental Management*, **88**, 1161-1174. http://dx.doi.org/10.1016/j.jenvman.2007.06.014

[17] Riley, T. and Barr, L. (2008) Partners in Protecting Arkansas' Water Bodies. University of Arkansas, Little Rock.

[18] Hoag, D.L.K., Chaubey, I., Popp, J., Gitau, M., Chang, L., Pennington, J., Rodríguez, H.G., Gbur, E., Nelson, M. and Sharpley, A.N. (2012) Lincoln Lake Watershed, Arkansas: National Institute of Food and Agriculture—Conservation Effects Assessment Project. In: Osmond, D.L., Meals, D.W., Hoag, D.L.K. and Arabi, M., Eds., *How to Build Better Agricultural Conservation Programs to Protect Water Quality*, Soil and Water Conservation Society, Ankeny, 171-200.

[19] Chiang, L., Chaubey, I., Gitau, M. and Arnold, J.G. (2010) Differentiating Impacts of Land Use Changes from Pasture Management in a CEAP Watershed Using the SWAT Model. *Transactions of the American Society of Agricultural Engineers*, **53**, 1569-1584.

[20] Chaubey, I., Chiang, L., Gitau, M. and Mohamed, S. (2010) Effectiveness of BMPs in Improving Water Quality in a Pasture Dominated Watershed. *Journal of Soil and Water Conservation*, **65**, 424-437. http://dx.doi.org/10.2489/jswc.65.6.424

[21] Gitau, M., Chaubey, I., Gbur, E., Pennington, J. and Gorham, B. (2010) Impact of Land Use Change and BMP Implementation on Water Quality in a Pastured Northwest Arkansas Watershed. *Journal of Soil and Water Conservation*, **65**, 353-368. http://dx.doi.org/10.2489/jswc.65.6.353

[22] Vendrell, P.F., Steele, K.F., Nelson, M.A. and Cash, W.L. (2000) Extended Water Quality Monitoring of the Lincoln Lake Watershed. *Arkansas Water Resources Center Report MSC*-296, University of Arkansas, Fayetteville.

[23] Nelson, M.A., Cash, L.W. and Trost, G.K. (2004) Water Quality Monitoring of Moores Creek above Lincoln Lake 2003. *Arkansas Water Resources Center Report MSC*-319, University of Arkansas, Fayetteville.

[24] Bailey, B.W., Haggard, B.E. and Massey, L.B. (2012) Water Quality Trends across Select 319 Monitoring Sites in Northwest Arkansas. *Arkansas Water Resources Center Report MSC*-365, University of Arkansas, Fayetteville.

12

Stable Oxygen and Deuterium Isotope Techniques to Identify Plant Water Sources

M. Edwin S. Lubis[1], I. Yani Harahap[1], Taufiq C. Hidayat[1], Y. Pangaribuan[1], Edy S. Sutarta[1], Zaharah A. Rahman[2], Christopher Teh[2], M. M. Hanafi[2]

[1]Indonesia Oil Palm Research Institute, Medan, Indonesia
[2]Universiti Putra Malaysia, Serdang, Malaysia
Email: e19m20@yahoo.co.id

Abstract

There is still very little information on the sources of water absorbed by oil palm plant. This information is very important for water management system in oil palm plantation. Thus, this study was carried out to determine current water sources absorbed by the oil palm roots using oxygen ($\delta^{18}O$) and deuterium isotopes (δD) techniques. Sketches of oxygen and deuterium isotope were total rainfall, throughfall, runoff, measurement at 5 soil depths (namely: 20 cm, 50 cm, 100 cm, 150 cm, and 200 cm), and oil palm stem. Results of this study showed huge variance in the values of oxygen and deuterium isotope. Based on Least Significant Difference (LSD) test, there was no significant value in the oxygen and deuterium isotope of stem water and others; however, a similar value was obtained at the depths of 0 - 20 cm and 20 - 50 cm with the stem water. This indicated that oil palm absorbed water from 0 - 50 cm depth. This result agreed with the oil palm rooting system, which has verified that the root quarter is the most active root of oil palm.

Keywords

Deuterium Isotope Technique, Plant Water Sources, Stable Oxygen

1. Introduction

Stable isotope has been developed as a powerful tool for investigating processes in plant-water relations such as recognizing plant water use and responding to different types of water sources, as well as acquiring better understanding of water utilization processes, water use efficiency, pattern, mechanism, and the ability to adopt in arid environments. Plants have to cope with various water sources such as rainwater, soil water, groundwater, sea water and the mixtures of these. Plant water sources are usually characterized by different isotopic signatures

($^{18}O/^{16}O$ and D/H ratios). At present, the measurement of δD, $δ^{18}O$ compositions of various potential water sources and stem water has become a significant means to identify plant water sources [1].

The analyses of hydrogen and oxygen stable isotope provide an effective approach for studying root water uptake. Zimmermann *et al.* [2] discovered that there was no hydrogen and oxygen isotope fractionation during root water uptake. Thus, the water absorbed by plant roots can be considered as a mixture of water from different water sources. By comparing the hydrogen and oxygen isotopes in the water from plant xylem and probable water sources, the contributing proportions of different water sources to plants can therefore be confirmed [3] [4].

There are marked differences in the stable fracture due to various water cycles resulting in the distinct difference of δD and $δ^{18}O$ in various water bodies [5]-[7]. Variation in these sources of water can be as large as 200‰ in a single location [5], which makes it possible to determine plant water sources and to understand the utilization process of various potential water sources, mechanism, as well as co-existence and competition between the plants and neighbouring plants.

Three potential sources of natural water (precipitation, stream, and groundwater) have been found to vary widely in isotopic composition [8]. Since there is no isotope fractionation associated with the uptake of soil water by roots [1] [9], the δD and $δ^{18}O$ values of plant xylem water represent a weighted average of all soil water acquired by all functional roots. The soil depth from which xylem water originates can therefore be estimated by comparing the δD or $δ^{18}O$ of plant xylem water with that of the soil water from the different parts of the profile. The palm root system is a system of root fibres, with the roots to a depth of 1 m, but most are at the depth of 15 - 30 cm. Although groundwater surface (water table) is quite deep, active root system is generally located at 5 - 35 cm depth, while the tertiary root is at 10 - 30 cm depth [10].

2. Material and Method

The research was conducted in Kabun-Aliantan, Tandun (N: 00°27, 925', E: 100°49, 219') in Riau, Indonesia. For this purpose, field observation was carried out on the land with the *typic Hapludut* soil type that has a clay texture, located at 30 - 70 m above the sea level and flat-undulated physiographically. This research was conducted in a plot area of oil palm plantation measuring 50 m × 50 m which consisted of 32 trees. The selection of the trees in the middle of the plot area included as many as 3 trees of oil palm, patterned equilateral triangle as a medium for stem water sampling.

The stem water sampling was conducted on the stem micro sampler type 1906 DVC. The equipment was installed by inserting it into the stem using the first stem pipe end capillary drill which functions to absorb water from the inner wall of palm trunks. The vacuum soil sampler is operated by absorbing water that utilises a suction pipe at the end of the capillary (**Figure 1(a)**). Meanwhile, groundwater samples were obtained by using a PVC vacuum pipe for one hour with a water machine (**Figure 1(b)**).

Water sampling was conducted by positioning 5 different depths (20 cm, 50 cm, 100 cm, 150 cm, and 200 cm) around each tree in a circle, with three replications of 1.5 - 2 m from the tree. This water sampling was carried

(a)   (b)

Figure 1. (a) A micro sampler; (b) PVC vacuum pipe.

out nine times for observation. In addition, water sampling was also conducted for the rainfall, through fall and run-off samples. The water samples were then analysed for the deuterium and oxygen isotopes in each sample observation using a laser spectrometry. The measurement of the isotope ratios of ^{18}O and D Laser Spectroscopy used a device equipped with a microprocessor control analyser with a precision stable isotope ratio analysis. This stable isotope ratio was determined according to the relative difference of a heavier isotope to the lighter isotope (which has greater abundance) as $^{18}O/^{16}O$ or D/H. From the analysis of the deuterium content (δ^2H or δD) and oxygen-18 ($\delta^{18}O$), stable isotopes of water molecules were obtained from information on the processes that had been experienced by the sample water as evaporation or mixing process between two sources of water.

The composition tools included LGR DLT-100 or LWIA (Liquid Water Analyser isotopes), comprising the analysis of the laser system and internal computer, a CTC LC-PAL liquid auto sampler, a small membrane vacuum pump, and an air chamber output channel that passes air through a column of Diorite for removal of moisture. The isotope ratio relative was calculated to that of a standard Mean Ocean Water (SMOW) as astandard Dawson [11]. The deuterium isotope calculation with V-SMOW is as follows:

$$\delta D = \left[\frac{D/H_{sample}}{D/H_{standard}} - 1 \right] \times 1000\%$$

The oxygen isotope calculation with V-SMOW is as follows:

$$\delta^{18}O = \left[\frac{^{18}O/^{16}O_{sample}}{^{18}O/^{16}O_{standard}} - 1 \right] \times 1000\%$$

All data were analysed using the statistical LSD test to compare the contents of deuterium and oxygen isotopes in the oil palm trunk and groundwater at each depth. The analysis of oil palm rooting was carried out at the sites to determine the layer that contains the active palm roots which absorb water sources for growth and development. The method of root sampling was done by making the soil profile with a depth of 0 - 200 cm, where the distance sampling was 2, 50 m from the tree trunk, which was then performed on various soil sampling depths (0 - 20 cm, 20 - 50 cm, 50 - 100 cm, 100 - 150 cm, and 150 - 200 cm) using a ring of soil samples, where the volume of each soil sample was 20 cm × 20 cm × 20 cm. The weighing of soil samples was performed based on the weight of dry samples. This was then carried out based on the criteria for root classifications (primary, secondary, tertiary and quarter) at any depth.

3. Results and Discussion

3.1. Analysis of Deuterium Isotope (δD) of Water Sample

Figure 2 shows that on 17 November 2010, the deuterium (δD) values for rainfall, through fall was close to groundwater at 50 cm depth, with an effective rainfall of 15.90 mm. [8] [11] reported that the isotopic composition of groundwater is a weighted average of a long-term precipitation input. On 6 March 2011, the deuterium (δD) sample for run-off had a value close to the deuterium (δD) sample for stem water, with an effective rainfall of 30.0 mm. On 13 September 2010, the deuterium (δD) value for stem water samples was close to the groundwater sample at 20 cm depth, with effective rainfall of 27.40 mm. On 4 September 2010, the groundwater sample with 100 cm depth was close to the deuterium (δD) for stem water sample, with effective rainfall of 2.60 mm. On 24 October 2010, the deuterium values for stem water samples were close to 150 cm and 200 cm of groundwater depths, with effective rainfall of 14.70 mm. It is important to note that plants can utilize water from precipitation, soil water, run off (including melting), groundwater, fog, and condensate water [5] [12].

Figure 3 shows that the deuterium concentrations of stem water samples were at the closest point with the deuterium of the groundwater samples at 50 cm depth, followed by the deuterium of groundwater samples at 20 cm depth, the deuterium of groundwater samples at 150 cm depth, the deuterium of ground water samples at 100 cm depth, the deuterium of rainfall samples, the deuterium of through fall samples, the deuterium of groundwater samples at 200 cm death, and the deuterium of run-off samples. The analysis carried out for the hydrogen and oxygen stable isotopes provides an effective approach for studying root water uptake [2].

In **Table 1**, the highest concentration of deuterium in the water samples was obtained from the run-off amounting to −58.722‰, where as the lowest concentration of deuterium was presented in the through fall, with the value −52.133‰. Based on the LSD test, it can be seen that the concentration of deuterium (δD) in the stem

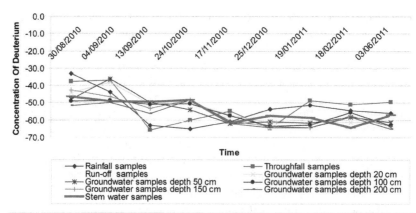

Figure 2. Deuterium isotope on the date of sampling time.

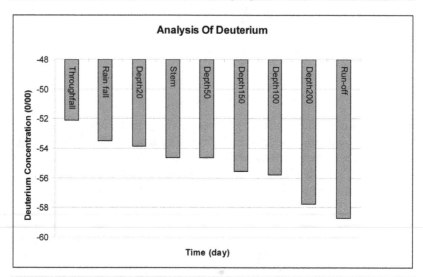

Figure 3. Deuterium concentrations of groundwater sample.

Table 1. Deuterium (δD) of water samples.

No.	Treatment	Deuterium Concentration (0/00)	
1	Water Sampling Rainfall	−53.478	A
2	Water Sampling Throughfall	−52.133	A
3	Water Sampling Run-Off	−58.722	A
4	Water Sampling Stem	−54.633	A
5	Water Sampling Depth 20 cm	−53.878	A
6	Water Sampling Depth 50 cm	−54.644	A
7	Water Sampling Depth 100 cm	−55.789	A
8	Water Sampling Depth 150 cm	−55.578	A
9	Water Sampling Depth 200 cm	−57.767	A

water samples (−54.633‰) approximated the deuterium concentration of water samples at 20 cm death with −53.878 ‰ and the deuterium concentration of water samples at 50 cm death with −54.644‰. It can be concluded that the value of the deuterium (δD) ratio in all types of treatment in this research did not provide a significantly different effect, whereby the deuterium data sorting concentrations ranged from highest to lowest. The measurement of δD, δ^{18}O composition of various potential water sources and stem water samples was a significant means to identify plant water sources [13].

3.2. Analysis of Oxygen Isotope (δ^{18}O) of the Water Samples

Figure 4 shows that the water sampling on 25 December 2010 yielded the oxygen isotope (δ^{18}O) values of stem water samples close to the oxygen isotope (δ^{18}O) values of the water samples for rainfall and groundwater sampling at 50 cm depth, with effective rainfall of 45.20 mm. On 17 September 2010, the values of oxygen isotope (δ^{18}O) for the through fall samples were close to the oxygen isotope (δ^{18}O) in the stem water samples, with effective rainfall of 15.90 mm. On 18 February 2011, the oxygen isotope (δ^{18}O) values for the run-off water samples showed that the oxygen isotope (δ^{18}O) values were similar to the oxygen isotope (δ^{18}O) values of the stem water samples in rainy conditions. On 24 October 2010, the samples of groundwater at the depths of 20 cm and 200 cm revealed the oxygen isotope (δ^{18}O) values which were close to the oxygen isotopic (δ^{18}O) values of the stem water samples, with effective rainfall of 14.70 mm. From **Figure 5**, it can be observed that the samples of groundwater, with the depths of 100 cm and 150 cm, showed the oxygen isotope (δ^{18}O) values that are similar to the oxygen isotope (δ^{18}O) values of the stem water samples, with effective rainfall of 27.40 mm.

Zimmermann *et al.* [2] discovered no hydrogen and oxygen isotope fractionation during root water uptake; thus, the water was absorbed by plant roots, which can be considered as a mixture of water from different water sources. The results for oxygen isotope (δ^{18}O) concentrations of groundwater sample analysis are displayed in **Figure 6**.

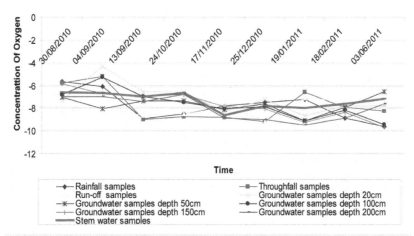

Figure 4. Oxygen isotope on the date of sampling time.

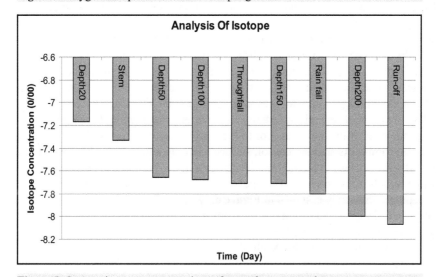

Figure 5. Oxygen isotope concentrations of ground water samples.

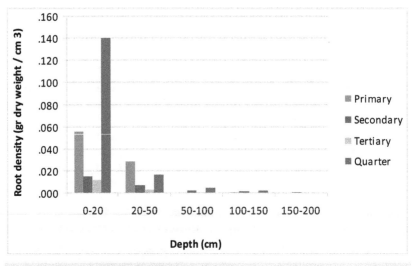

Figure 6. Palm root density in different soil depths.

From **Figure 5** above, the concentrations of oxygen isotope of the stem water samples were at the closest point to the groundwater samples at 20 cm depth, followed by the oxygen isotope samples of groundwater at 50 cm depth, the oxygen isotope samples of groundwater at 100 cm depth, the oxygen isotope of throughfall water samples, the oxygen isotope samples of groundwater at 150 cm depth, the oxygen isotope samples of rainfall, the oxygen isotope samples of groundwater at 200 cm depth, and the oxygen isotope samples of run-off. According to Allison [14], the oxygen of the upper soil water (at 0 - 20 cm depth) was always more enriched due to evaporation than the lower soil horizons and groundwater.

Table 2 reveals that the highest concentration of oxygen isotopes ($\delta^{18}O$) was obtained from the run-off water sample, *i.e.* −8.067‰. On the other hand, the lowest concentration of oxygen isotopes was found in the water samples at 20 cm depth, with the value −7.171‰. Based on the LSD test, the concentration of oxygen isotopes ($\delta^{18}O$) in the stemwater samples (−7.334‰) approximated the concentration of oxygen isotope at 20 cm depth of the water samples and the concentration of −7.171‰ of the oxygen isotope samples at 50 cm depth, with the value −7.663‰. Therefore, it can be concluded that palm oil plant sabsorb more dominant sources of water at the depths of 0 - 20 cm and 20 - 50 cm for growth and development, respectively. Walker & Richardson [1] stated that by comparing the isotopic composition of water between soil and plant, water uptake can be concluded to have occurred at a depth of 20 to 60 cm, which is in good agreement with the root and soil water potential distributions.

From **Table 3** above, the highest mean value in the quarter root and the lowest root in the tertiary can be seen at the depth of 0 - 20. At the depth of 20 - 50 cm, the highest value rates were found in the primary root, where as the lowest values were obtained in the tertiary. At the depth of 50 - 100 cm, the highest value was found in the quarter root. At the depth of 100 - 150, the highest rate was found in the root quarter, while the lowest was found in the primary root. In addition, the average depth of 150 - 200 contained the highest value on the secondary root, where as the lowest value was found in the primary and quarter roots. Based on the information in **Table 3** it can be concluded that the rooting of oil palm is more dominant at the depths of 0 - 20 and 20 - 50 cm. According to Fauzi *et al.* [14], the roots of secondary, tertiary, and quarter grow parallel to the ground surface roots and even to the tertiary and quarter of the upper layer or to the places that contain lots of nutrients. Furthermore, Iyung [4] stated that most oil palm roots are near the soil surface and only a few roots of palm oil are at the depth of 90 cm. Although ground water surface (watertable) is quite deep, active root system is generally located at the depth of 5 - 35 cm, while tertiary roots are located at the depth of 10 - 30 cm. The result analysis of the root samples at different soil depths in this study is shown in **Figure 6**.

Figure 6 clearly shows that the rooting of oil palm that is dominated by the quarter root at 0 - 20 cm depth. Hartley [15] stated that the rooting system of palm oil is the root system of fibres, with the roots to a depth of 1 m. However, most roots are found at the depth of 15 - 30 cm. The root system consists of the upper roots of primary, secondary, tertiary, and quarter. The root of the current quarter is the root that absorbs nutrients, water and oxygen.

Table 2. Oxygen isotopes of water samples.

No.	Treatment	Deuterium Concentration (0/00)	
1	Water Sampling Rainfall	−7.796	A
2	Water Sampling Throughfall	−7.708	A
3	Water Sampling Run-Off	−8.067	A
4	Water Sampling Stem	−7.334	A
5	Water Sampling Depth 20 cm	−7.171	A
6	Water Sampling Depth 50 cm	−7.663	A
7	Water Sampling Depth 100 cm	−7.682	A
8	Water Sampling Depth 150 cm	−7.713	A
9	Water Sampling Depth 200 cm	−7.997	A

Table 3. The root density by soil depth.

Depth (Cm)	Primary	Secondary	Tertiary	Quarter
	gr/cm^3			
	0.0151	0.0091	0.0065	0.1261
0 - 20	0.0548	0.0166	0.0124	0.1169
	0.0981	0.0206	0.0164	0.1766
Average	0.0560	0.0155	0.0118	0.1399
	0.0190	0.0044	0.0020	0.0161
20 - 50	0.0193	0.0035	0.0023	0.0135
	0.0483	0.0141	0.0055	0.0200
Average	0.0288	0.0073	0.0033	0.0165
	0.0	0.0	0.0009	0.0039
50 - 100	0.0	0.0028	0.0005	0.0046
	0.0	0.0049	0.0005	0.0055
Average	0.0000	0.0025	0.0006	0.0047
	0.0	0.0	0.0020	0.0018
100 - 150	0.0	0.0029	0.0004	0.0019
	0.0014	0.0010	0.0001	0.0025
Average	0.0005	0.0013	0.0008	0.0020
	0.0	0.0	0.0004	0.0
150 - 200	0.0	0.0005	0.0004	0.0
	0.0	0.0011	0.0000	0.0
Average	0.0000	0.0005	0.0003	0.0000

4. Conclusion

The analysis of deuterium (δD) and oxygen isotopes ($\delta^{18}O$) in groundwater and water in oil palm trunks provides information on the dynamics of plant utilization of water resources. Based on the Least Significant Difference (LSD) test, no significant value was found in the deuterium and oxygen isotopes in the stem water samples and other samples. However, similar values were obtained at the depths of 0 - 20 cm and 20 - 50 cm in the stem water. It indicates that oil palm absorbs water from the depth of 0 - 50 cm. This result is in accordance with the system of oil palm rooting, *i.e.* the root quarter (0 - 50 cm) is the most active root of oil palm that absorbs nutrients, water and oxygen.

References

[1]   Walker, G.R. and Richardson, S.B. (1991) The Use of Stable Isotopes of Water in Characterizing the Source of Water in Vegetation. *Chemical Geology (Isotope Geoscience Section)*, **94**, 145-158. http://dx.doi.org/10.1016/0168-9622(91)90007-J

[2]   Zimmerman, U., Ehhalt, D. and Munnich, K.O. (1967) Soil Water Movement and Evapotranspiration: Changes in the Isotopic Composition of the Water. In: *Isotopes in Hydrology, Proceedings of the Symposium*, International Atomic Energy Agency (I.A.E.A.), Vienna, 567-584.

[3]   Brunel, J.P., Walker, G.R. and Keennett-Smith, A.K. (1995) Field Validation of Isotopic Procedures for Determining Source Water Used by Plants in Semi-Arid Environment. *Journal of Hydrology*, **167**, 351-368. http://dx.doi.org/10.1016/0022-1694(94)02575-V

[4]   Iyung, P. (2007) Palm Complete Guide Agribusiness Management from Upstream to Downstream. Self-Spreader, Jakarta.

[5]   Corbin, J.D., Thomsen, M.A., Dawson, T.E. and D'Antonio, C.M. (2005) Summer Water Use by California Coastal Prairie Grasses: Fog, Drought, and Community Composition. *Oecologia*, **145**, 511-521. http://dx.doi.org/10.1007/s00442-005-0152-y

[6]   Dansgaard, W. (1964) Stable Isotope in Precipitation. *Tellus*, **16**, 436-468. http://dx.doi.org/10.1111/j.2153-3490.1964.tb00181.x

[7]   Ingraham, N.L. and Taylor, B.E. (1991) Light Stable Isotope Systematic of Large-Scale Hydrologic Regimes in California and Nevada. *Water Resources Research*, **27**, 77-90. http://dx.doi.org/10.1029/90WR01708

[8]   Ehleringer, J.R. and Dawson, T.E. (1992) Water-Uptake by Plants: Perspectives from Stable Isotope Composition. *Plant, Cell and Environment*, **15**, 1073-1082. http://dx.doi.org/10.1111/j.1365-3040.1992.tb01657.x

[9]   Förstel, H. (1982) $^{18}O/^{16}O$ Ratio of Water in Plants and Their Environment. In Schmidt, H.L., Förstel, H. and Heinzinger, K., Eds., *Stable Isotopes*, Elsevier, Amsterdam, 503-516.

[10] Yang, Q., Xiao, H., Zhao, L., Zhou, M., Li, C. and Cao, S. (2010) Stable Isotope Techniques in Plant Water Sources: A Review. *Sciences in Cold and Arid Regions*, **2**, 112-122.

[11] Dawson, T.E. and Pate, J.S. (1996) Seasonal Water Uptake and Movement in Root Systems of Australian Phraeatophytic Plants of Dimorphic Root Morphology: A Stable Isotope Investigation. *Oecologia*, **107**, 13-20. http://dx.doi.org/10.1007/BF00582230

[12] Jackson, R.B., Canadell, J., Ehleringer, J.R., Mooney, H.A., Sala, O.E. and Schulze, E.D. (1996) A Global Analysis of Root Distributions for Terrestrial Biomes. *Oecologia*, **108**, 389-411. http://dx.doi.org/10.1007/BF00333714

[13] Allison, G.D. and Hughes, M.W. (1983) The Use of Natural Tracers as Indicators of Soil-Water Movement in a Temperate Semi-Arid Region. *Journal of Hydrology*, **60**, 157-173. http://dx.doi.org/10.1016/0022-1694(83)90019-7

[14] Fauzi, Y. (2008) Palm Cultivation & Utilization of Waste Analysis Business & Marketing. Revised Edition, Self-Spreader, Jakarta.

[15] Hartley, C.W.S. (1977) The Palm Tree (*Elaeis guineensis* Jacq.). Longman, London.

13

Evaluating Water Stability Indices from Water Treatment Plants in Baghdad City

Awatif S. Alsaqqar, Basim H. Khudair, Sura Kareem Ali

Department of Civil Engineering, College of Engineering, Baghdad University, Baghdad, Iraq
Email: d.alsaqqar@yahoo.com, basim22003@yahoo.com, srak39@yahoo.com

Abstract

Corrosion control is an important aspect of safe drinking water supplies. The effects of corrosion which may not be evident without monitoring are an important issue concerning both public health and economical aspects. Chemical stability parameters of water quality in water treatment plants in Baghdad city can improve drinking water quality. The treated water quality from water treatment plants in Baghdad city was investigated along the water flow path in this study. The water quality parameters related to chemical stability included temperature, alkalinity as mg/L CaCO$_3$, calcium mg/L as Ca, pH and total dissolved solids (TDS) mg/L for different samples from WTPs within Baghdad city were investigated. The two water quality indices, Langelier saturation index (LSI) and the Ryznar stability index (RSI), were calculated in order to evaluate the chemical stability of the drinking water samples. The results of LSI and RSI of the effluents from Baghdad's WTPs during 2000-2013 classified that corrosive water is produced and this indicates that the water is not safe for domestic use and will need the further treatment. The present study demonstrated the application of water stability indices in estimating/understanding the treated water chemical stability and appeared to be promising in the field of treated water quality management.

Keywords

Water Stability Indices, Chemical Stability, Langelier Index, Ryznar Index

1. Introduction

1.1. Water Stability

Stability of water is the tendency of water to either dissolve or deposit minerals varying with its chemical makeup. Water that tends to dissolve minerals is considered corrosive and that tends to deposit mineral is considered scaling. Corrosive water can dissolve minerals like calcium and magnesium, also can dissolve harmful metals such as lead and copper from plumbing utilities. Where scaling waters deposit a film of minerals on pipe walls and may prevents corrosion of metallic surfaces. If the scale deposition is too rapid, it also can be harmful and can damage appliances, such as water heaters, and increase pipe friction coefficients, in extreme cases, scale may clog pipes. Therefore, the most desirable water is one that is just slightly scaling [1].

Control of water quality in the distribution and plumbing systems seeks to preserve the basic characteristics of water during its conveyance from the point of production and treatment to the consumers tap. The finished water should be completely stable in its compositional and physical attributes. Also the conveyance system and accessory structures (pipelines, distributing reservoirs, mains and serves pipes) should be reactively inert to the water being conveyed. Most water quality parameters affect the corrosion process to some degree and that each pipe material is affected differently. Altering one parameter to subdue its effect on the corrosively of water may well change other water quality characteristics, perhaps rendering the water even more corrosive or less likely in some other way to meet drinking water standards [2]. At present, there is no requirement to produce water that is stable (either scale forming or corrosive) other than having a pH in the range 6.5 to 8.5 [3].

Water quality changes in drinking water distribution systems occur as a result of complex and often interrelated chemical and biological processes [4]. The chemical instability of water quality shows the occurrence of scale deposition and corrosion which can cause secondary pollution of water quality, increase energy consumption of water transportation and decrease service life of pipe networks [5].

Corrosive water is related to its pH, alkalinity, hardness, temperature, dissolved oxygen, carbon dioxide, total dissolved solids and other physical, chemical and biological factors. Water with high levels of sodium, chloride, or other ions will increase the conductivity of the water causing corrosion and also be accelerated by high: flow rates within the piping system, water temperature, dissolved O_2 and CO_2 and dissolved salts such as sulfates [6].

1.2. Water Stability Indices

Although a number of indices have been developed, none has demonstrated the ability to accurately quantify and predict the corrosively or scaling of water. They can only give a probable indication. Experience has shown that if conditions encourage the formation of a protective calcium carbonate film, then corrosion will generally be minimized [7]. The most common methods used for calculating the stability of water are Langelier saturation index (LSI) and Ryznar stability index (RSI). LSI and RSI are designed to be predictive tools for calcium carbonate scale and they are not suitable for estimating calcium phosphate, calcium sulfate, silica or magnesium silicate scales [8]. Millete *et al.* [9] showed that almost 70% of the representative utilities to have moderately to highly aggressive waters, where 16% - 18% had highly aggressive waters and indicating aggressive to very aggressive waters as well as a comparison between the two indices showed good agreement in the results.

Pisigan and Singly [10] reported the influence of some water quality parameters in particular the LSI on the corrosion of galvanized steel using a dynamic circulating water system. Change in water quality parameter were monitored, hardness and alkalinity exhibited a continuous decline. Pisigan and Singly, [11] dictated from their experimental work that increasing the buffer capacity at a constant water alkalinity of 100 mg/L as $CaCO_3$ in the pH range 6 - 9 decreased the corrosion rate of mild steel. However an increase in the buffer capacity at various pH by raising the alkalinity did not lower the corrosion rate because of the effects of higher ionic strength and conductivity. Also higher chloride content increased the corrosion rate in some waters of high buffer capacities. Steve *et al.* [12] evaluated the relationship between copper plumbing corrosion and variations in delivered water quality in several communities in the Pacific Northwest. Significant relationships were found for copper corrosion rate dependence on pH and free chlorine residual. The regression equation coefficient indicated that the corrosion rate changes with chlorine residual, approximately 25.6 μm/year per mg/L chlorine.

1.3. Research Objective

This paper is to evaluate the stability of the water flowing in the distribution systems in Baghdad city from the existing water treatment plants. Two indices may be calculated to assess the corrosively of water due to problems associated with calcium carbonate scale formation. Probably the best way to qualitatively predict the formation of $CaCO_3$ scale is through the use of the Langelier Saturation Index (LSI) and the Ryznar Stability index

(RSI). The calculation of these indices is through the investigation of the chemical stability parameters including Temperature, Alkalinity as $CaCO_3$, Calcium as Ca, pH and total dissolved solids (TDS) for the effluents from the WTPs within Baghdad city.

2. Materials and Methods

2.1. Study Area Description

Tigris river water is considered the only source of potable water for the city of Baghdad, and the river divides the city into right (Karkh) and left (Risafa) sides with a flow direction from north to south. The study area within Baghdad City is located in the Mesopotamian alluvial plain between latitudes 33°14'N - 33°25'N and longitudes 44°31'E - 44°17'E, 30.5 to 34.85 m at sea level (a.s.l). The area is characterized by arid to semi-arid climate with dry hot summers and cold winters; the mean annual rainfall is about 151.8 mm [13]. In Baghdad city, a tremendous increase in freshwater demand is required due to the rapid growth in population and accelerated industrialization. The quality of the following water is affected by the pollution increase in the river stretch due to effluent discharges by various uncontrolled sources as domestic, industries, agriculture along the downstream stretch. Therefore treated water quality from WTPs monitoring is necessary to evaluate the treated water stability.

2.2. Data Collection and Analysis

All water treatment plants (WTPs) in Iraq are designed as conventional plants. This treatment process does not significantly affect the concentrations of the dissolved constituents, so the characteristics of the raw and treated water are quite the same. There are eight water treatment plants (WTPs) from the north to the south of Baghdad city, Al-Karkh, East Tigris, Al-Wathbah, Al-Karamah, Al-Qadisiya, Al-Dawrah, Al-Rashid, Al-Wahda WTPs (**Figure 1**). The data used in this study were sampled, tested and applied from Baghdad Mayoralty (Amanat

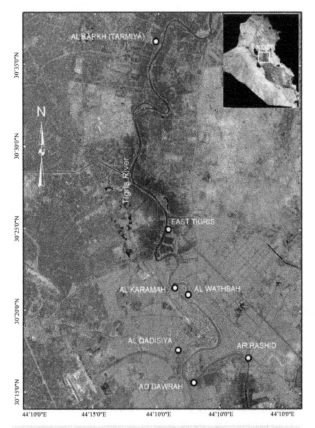

Figure 1. Sampling locations across Tigris River, within Baghdad City. Location of water treatment plant.

Baghdad)-Water office which represents the quality of the treated water from the WTPs in Baghdad city according to WHO standards recorded for the period between 2000 until 2013. In order to evaluate the chemical stability of any type of water, many parameters are detected, such as Temperature, Alkalinity mg/l as CaCO₃, pH, Total dissolved Solids (TDS), and concentration of Calcium (Ca). The variation in water stability with time and distance from the north to the south of Baghdad is to be indicated.

2.3. Calculations of the Water Stability Indices

2.3.1. Langelier Saturation Index

Work by professor W. F. Langelier, published in 1936 deals with the conditions at which water is in equilibrium with calcium carbonate. An equation developed by Langelier makes it possible to predict the tendency of calcium carbonate either to precipitate or to dissolve under varying conditions. Langelier Saturation Index (LSI) is an equilibrium model derived from the theoretical concept of saturation and provides an indicator of the degree of saturation of water with respect to calcium carbonate and can calculate from [14]:

$$LSI = pHa - pHs \tag{1}$$

where:

pHa: the measured water pH.

pHs: the pH at which water with a given calcium content and alkalinity is in equilibrium with calcium carbonate.

The equation expresses the relationship of: pH, calcium, total alkalinity, dissolved solids, and temperature as they are related to the solubility of calcium carbonate in waters with pH of 6.5 to 9.5. This is known as the pHs:

$$pHs = (9.3 + A + B) - (C + D) \tag{2}$$

where:

$$A = \frac{\left[\text{Log}_{10}(\text{TDS}) - 1 \right]}{10} \tag{3}$$

$$B = -13.12 \times \text{Log}_{10}(^{\circ}C + 273) + 34.55 \tag{4}$$

$$C = \text{Log}_{10}\left(Ca^{+2} \text{ as } CaCO_3\right) - 0.4 \tag{5}$$

$$D = \text{Log}_{10}\left[\text{Alkalinityas } CaCO_3\right] \tag{6}$$

2.3.2. Ryznar Stability Index (RSI)

The Ryznar index is an empirical method for predicting scaling tendencies of water based on study of operating results with water at various saturation indices. The Stability Index developed by John Ryzner in 1944 used the Langelier Index (LSI) as a component in a new formula to improve the accuracy in predicting the scaling or corrosion tendencies of water [14].

The Ryznar index (RSI) takes the form:

$$(RSI) = 2(pHs) - pH = pHs - LSI \tag{7}$$

The criteria used to give an indication of the stability indices can be summarized in **Table 1**.

3. Results and Discussion

3.1. Treated Water Quality

Water quality parameters from Baghdad's WTPs compared with Iraqi and WHO drinking water standards are shown in **Table 2**, where the treated water is within these standards. In order to reach a better view on the variation of the treated water quality from the WTPs and how they may affect water stability, selected results from the determination of water quality parameters are discussed below.

Alkalinity in water is due to the presence of weak acid systems that consume hydrogen ions produced by other reactions or produce hydrogen ions when they are needed by other reactions allowing chemical or biological

Table 1. Summary of water stability indices [18].

Index value	Water condition
LSI > 0	Water is supersaturated with respect to calcium carbonate ($CaCO_3$) and scale forming and $CaCO_3$ precipitation may occur.
LSI = 0	Water is considered to be neutral. Neither scale-forming nor scale removing. Saturated, $CaCO_3$ is in equilibrium. Borderline scale potential.
LSI < 0	Water is under saturated with respect to calcium carbonate. Under saturated water has a tendency to remove existing calcium carbonate protective coatings in pipelines and equipment. No potential to scale, the water will dissolve $CaCO_3$.
RSI ≤ 6	Supersaturated, tend to precipitate $CaCO_3$. The scale tendency increase as the index decrease.
6 < RSI < 7	Saturated, $CaCO_3$ is in equilibrium. The calcium carbonate formation probably does not lead to a protective corrosion inhibitor film.
LSI ≥ 7	Under saturated, tend to dissolved $CaCO_3$. Mild steel corrosion becomes an increasing problem.

Table 2. Statistical summary of Baghdad's WTPs-treated water quality data.

Parameter	Min.	Max.	Average	Drinking standards	
				WHO	Iraqi
Temperature (°C)	21	22.81	21.89	-	-
Alkalinity as $CaCO_3$ (mg/L)	124.08	143	137.77	30 - 500	125 - 200
Calcium as Ca (mg/L)	70	120.75	83.46	75	75
pH	7.45	7.90	7.63	7 - 8.5	6.5 - 8.5
Total dissolved Solids TDS (mg/L)	410	696.5	541.53	500	1000

activities to take place within water without changing the pH [15]. The primary source of alkalinity is the carbonate system, although phosphates, silicates, borates, carboxylates, and other weak acid systems can also contribute. In water treatment plants, alkalinity is required in the coagulation process for the reaction of alum; lime could be added if the natural alkalinity is not enough for this reaction [8]. From the recorded data the variation in alkalinity ranged between 124.08 to 143 mg/L with average 137.77 mg/L as $CaCO_3$ as shown in **Table 2**. Corrosivity of water decreases as the alkalinity increases where it is necessary to provide a stable pH throughout the distribution system for corrosion control of metal pipes [16]. Low iron corrosion rates and iron concentrations in the distribution systems have been associated with higher alkalinities [11].

Calcium is the second most prevalent constituent in most surface waters and is generally among the most prevalent three or four ions in groundwater. Weathering and soil ion exchange reactions are the main sources of calcium in natural waters [8]. Increasing the Ca concentration will decrease the corrosivity of water, as Ca is important in various roles, including calcium carbonate scales, mixed iron/calcium carbonate solids and the formation of a passivating film on the surface of the pipe, to control corrosion [16]. From the recorded data the concentration of Ca ranged between 70 to 120.75 mg/L with an average of 83.46 mg/L in all WTPs as shown in **Table 2**.

pH is important in water treatment as it directly influences the dosages of chemicals added to reduce hardness and coagulate particles [8]. High pH of the water flowing in the distribution system will decrease the solubility of the corrosion by products formed in the system [16]. The results of pH varied from 7.45 to 7.90 with average 7.63, indicating that the water is almost neutral to sub-alkaline in nature.

Total dissolved solids (TDS) in water are due to inorganic salts. Principally these inorganic constituents are calcium, magnesium, sodium, and potassium salts of bicarbonate, chloride, sulfate, nitrate and phosphate. These compounds originate from weathering and leaching of rocks, soils and sediments. Also some of these compounds are added to the water in the treatment plants [8]. From the recorded data the variation concentration of TDS ranged between 410 to 696.5 mg/L with an average of 541.53 mg/L as shown in **Table 2**. The conventional treatment process does not affect the dissolved content of the treated water so the high TDS concentrations in the effluent are due to the raw water quality entering these plants. The increasing amounts of dissolved solids in the Tigris River are due to the discharge of waste water, leaching of fertilizers and the natural weathering of the soil. The effect of TDS content on water corrosivity is a complex issue. Some species such as carbonate and bicarbonate reduce corrosion, whereas chloride, sulfate, bromide and nitrate ions markedly accelerate corrosion [5].

Temperature also affects the corrosion process. Higher water temperatures accelerate the rate of corrosion by increasing the rate of the cathodic reaction. The chemical reaction rate generally is doubled for every 8°C increase in temperature [17] [18].

3.2. Langelier vs Ryzner Stability Indices

The stability of the treated water from all WTPs in Baghdad was calculated according to the water quality of the effluents of these plants. The US Environmental Protection Agency (USEPA) has recommended the use of Langelier (LSI) and Ryznar (RSI) stability indexes by utilities to monitor the corrosion potential of water [10]. **Table 3** shows that the correlation coefficient between LSI and the water quality parameters, it is observed that pH had the highest significant relationship with R = 0.507 at ($P < 0.01$) and then temperature (R = 0.421) while the other parameters had a weak relationship. Temperature had the highest significant relationship with RSI (R = 0.468) as shown in **Table 4**, while alkalinity and pH came second with R = 0.306 and R = 0.311 respectively.

Table 5 shows the stability indices LSI and RSI of the effluents from Baghdad's WTPs. The average annual LSI results are classified as positive values in East Tigris, Al-Wathbah, Al-Karamah, Al-Qadisiya, and Al-Rashid WTPs which refer to water is supersaturated with respect to calcium carbonate ($CaCO_3$) where scale forming of $CaCO_3$ precipitation may occur. But negative values were calculated at Al-Karkh, Al-Dawrah, and Al-Wahda WTPs which refer that the water is under saturated with respect to calcium carbonate. Under saturated water has the tendency to remove existing calcium carbonate protective coatings in pipelines and equipment. While average annual RSI results in all WTPs can be classified more than 7 and ranged between 7.344 to7.785 which refers to under saturated, or classified corrosive water as indicated in **Table 1**.

3.3. Annual LSI & RSI

The annual LSI and RSI for the period 2000-2013; are shown in **Table 6**. LSI results show that, for treated water

Table 3. Correlation coefficient between LSI and treated water quality parameters.

	LSI	Temp	Alkalinity	Ca	pH	TDS
LSI	1					
Temp	0.424(**)	1				
Alkalinity	−0.221(*)	−0.163	1			
Ca	0.286(**)	−0.119	−0.553(**)	1		
pH	0.507(**)	0.381(**)	0.227(*)	0.198(*)	1	
TDS	0.120	−0.235(**)	−0.447(**)	0.925(**)	0.129	1

**Correlation is significant at the 0.01 level (2-tailed). *Correlation is significant at the 0.05 level (2-tailed).

Table 4. Correlation coefficient between RSI and treated water quality parameters.

	RSI	Temp	Alkalinity	Ca	pH	TDS
RSI	1					
Temp	−0.468(**)	1				
Alkalinity	0.306(**)	−0.163	1			
Ca	−0.280(**)	−0.119	−0.553(**)	1		
pH	−0.311(**)	0.381(**)	0.227(*)	0.198(*)	1	
TDS	−0.127	−0.235(**)	−0.447(**)	0.925(**)	0.129	1

**Correlation is significant at the 0.01 level (2-tailed). *Correlation is significant at the 0.05 level (2-tailed).

Table 5. Annual LSI and RSI of effluent water from WTPs from 2000-2013.

Year	Al-Karkh	East Tigris	Al-Wathbah	Al-Karamah	Al-Qadisiya	Al-Dawrah	Al-Rashid	Al-Wahda
LSI	−0.122	0.032	0.042	0.195	0.168	−0.006	0.045	−0.012
RSI	7.785	7.595	7.505	7.344	7.338	7.540	7.473	7.551

Table 6. Annual LSI and RSI of effluent water from WTPs from 2000-2013.

Year	2000	2001	2002	2003	2004	2005	2006	2007	2008	2009	2010	2011	2012	2013
LSI	0.156	0.074	0.067	0.081	0.002	0.147	0.071	0.370	0.120	−0.082	−0.136	−0.065	0.021	0.002
RSI	7.38	7.54	7.5	7.53	7.59	7.40	7.52	7.17	7.34	7.66	7.68	7.63	7.54	7.58

from year 2000-2008 and 2012-2013 are positive values indicating scale forming water while from 2009-2011 are negative values or corrosive water. While RSI results from period 2000-2013 show more than 7 as under saturated, tend to dissolved $CaCO_3$ or corrosive water.

The differences between LSI and RSI shown in **Table 5** and **Table 6** (indicating corrosive or scale forming) can be according to: LSI measures only the directional tendency or driving force for $CaCO_3$ to precipitate or dissolve. It cannot be used as a quantitative measure, as two different waters one with low hardness (corrosive) and the other of high hardness (scale forming) can have the same LSI. While RSI makes it possible to distinguish between the two types, as RSI is based on the actual operating results with waters having various saturation indices [14] so it is more reliable.

4. Conclusions

As the choice of any index may propose different description water stability, so selecting the most appropriate index conforming to the actual situation of any system is the most sophisticated stage in making predictions. According to the present study:

1) The treated water from the WTPs in Baghdad is considered corrosive. So these plants need to pay attention to achieve national water quality standards and reduce corrosion and corrosion by-products, with corrective action that should be made to prevent corrosion in water supply networks.

2) Water quality affects the corrosivity of the treated water flowing from the WTPs into the distribution system in Baghdad city.

3) The calculated stability indices LSI and RSI indicated that the treated water from the plants was corrosive.

4) Using LSI together with the RSI contributes to more accurate prediction of the scaling or corrosive tendencies of water.

5. Recommendations

Control of water quality in the distribution system seeks to preserve the basic characteristics of water during its conveyance from the point of production and treatment to the consumers tap. The finished water from the treatment plants should be completely stable in its compositional and physical attributes. Several methods could be applied in the treatment plants to produce stable water like pH adjustment or adding corrosion inhibitors.

Acknowledgements

Our special gratitude and many thanks are forwarded to "Baghdad Mayoralty-Water Office" for supplying the data and without their support and encouragement the work would have not been done.

References

[1] Qasim, S.R., Edward, M.M. and Guany, Z. (2000) Planning, Design, and Operation. Prentice Hall PTR. Upper Saddle River, NJ07458.

[2] AWWA (1971) Water Quality and Treatment. 3rd Edition, McGraw Hill Book Company, New York.

[3] Australian Drinking Water Guidelines (1996) National Water Quality Management Strategy. NHMRC and ARMCANZ, Canberra.

[4] Imran, S.A., Dietz, J.D., Mutoti, G., Taylor, J.S., Randall, A.A. and Cooper, A.C.D. (2005) Red Water Release in Drinking Water Distribution System. *American Water Works Association*, **97**, 93-100.

[5] Fang, W. (2004) The Research on Water Chemical Stabilization and Control Methods in the Urban Water Supply System. Hunan University, Changsha.

[6] Salvato, J.A., Dee, P.E., Nemerow, N.L. and Agardy, F.J. (2003) Environmental Engineering. John Wiley & Sons, Inc., Hoboken.

[7] Rossum, J.R. and Merrill, D.T. (1983) An Evaluation of the Calcium Carbonate Saturation Indexes. *Journal of American Water Works Association*, **75**, 95-100.

[8] MWH (2005) Water Treatment Principles and Design. 2nd Edition, John Wiley & Sons, Inc., Hoboken.

[9] Millette, J.R., Hammonds, A.F., Pansing, M.F., Hansen, E.C. and Clark, P.J. (1980) Aggressive Water: Assessing the Extent of the Problem. *American Water Works Association*, **72**, 262-266.

[10] Pisigan Jr., R.A. and Singley, J.E. (1985) Effects of Quality Parameters on the Corrosion of Galvanized Steel. *American Water Works Association*, **77**, 76-82.

[11] Pisigan Jr., R.A. and Singley, J.E. (1987) Influence of Buffer Capacity, Chlorine Residual and Flow Rate on Corrosion of Mild Steel and Copper. *American Water Works Association*, **79**, 62-70.

[12] Reiber, S.H., Ferguson, J.F. and Benjamin, M.M. (1987) Corrosion Monitoring and Control in the Pacific Northwest. *American Water Works Association*, **79**, 71-74.

[13] Al-Adili, A.S. (1998) Geotechnical Evaluation of Baghdad Soil Subsidence and Their Treatments. Ph.D. Thesis, University of Baghdad, Baghdad.

[14] Water Service LTD (2004) Indexes for Calcium Carbonate. export@power-chemicals.com

[15] Gebbie, P. (2000) Water Stability: What Does It Mean and How Do You Measure It? *Proceedings of the 63th Annual Water Industry Engineers and Operators Conference*, Warrnambool, 6-7 September 2000, 50-58.

[16] Schock, M.R. (1989) Understanding Corrosion Control Strategies for Lead. *American Water Works Association*, **81**, 88-100.

[17] Kawamura, S. (2000) Integrated Design and Operation of Water Treatment Facilities. 2nd Edition, John Wiley & Sons, Inc., New York.

[18] Hadi, M. (2010) Development Software for Calculation of Eight Important Water Corrosion Indices. *Proceedings of the 12th National Conference on Environmental Health*, Tehran.

14

Water Resources of Uganda: An Assessment and Review

Francis N. W. Nsubuga[1,2]*, Edith N. Namutebi[3], Masoud Nsubuga-Ssenfuma[2]

[1]Department of Geography, Geoinformatics and Meteorology, University of Pretoria, Pretoria, South Africa
[2]National Environmental Consult Ltd., Kampala, Uganda
[3]Ministry of Foreign Affairs, Kampala, Uganda
Email: *nwasswa@gmail.com

Abstract

Water resources of a country constitute one of its vital assets that significantly contribute to the socio-economic development and poverty eradication. However, this resource is unevenly distributed in both time and space. The major source of water for these resources is direct rainfall, which is recently experiencing variability that threatens the distribution of resources and water availability in Uganda. The annual rainfall received in Uganda varies from 500 mm to 2800 mm, with an average of 1180 mm received in two main seasons. The spatial distribution of rainfall has resulted into a network of great rivers and lakes that possess big potential for development. These resources are being developed and depleted at a fast rate, a situation that requires assessment to establish present status of water resources in the country. The paper reviews the characteristics, availability, demand and importance of present day water resources in Uganda as well as describing the various issues, challenges and management of water resources of the country.

Keywords

Water Resources Management, Water Resources Utilisation, Climate Change, Water Resources Development, Uganda

1. Introduction

Of all the renewable resources, water has a unique place. It is essential in sustaining all form of life, food production, promoting economic development and for general wellbeing. Water is impossible to substitute for most of its uses, difficult to de-pollute, expensive to transport but a manageable natural resource capable of diversion,

---
*Corresponding author.

transport, storage and recycling [1]. There are two distinct water resources categories (surface and ground water) which are part of the earth's hydrologic cycle. Surface and groundwater resources play a major role in domestic water supply, watering livestock, industrial operations, hydropower generation, agriculture, marine transport, fisheries, waste discharge, tourism, and environmental conservation. The water resources of a region, conceived as a dynamic phase of the hydrologic cycle are influenced by climatic, physiographic and geological factors [2]. Orography on the other hand, plays a significant role in influencing rainfall and other climatic elements (temperature, humidity and wind) that affect evapo-transpiration, which determines the totality of water resources [2].

Water resources in Uganda comprise of large lakes like; Lake Victoria, Kyoga, Albert, George and Edward; wetlands and rivers, such as the Nile River, Katonga, Semliki, Malaba; rainfall, surface water runoff and ground water [3]. Two or more of these water resources are shared by other riparian states that are part of the Nile Basin, shown in the (**Table 1**). In a regional context, the whole of Uganda's water resources is part of the Nile. Uganda is a downstream riparian to Burundi, Democratic Republic of Congo (DRC), Kenya, Tanzania and Rwanda and an upstream riparian to South Sudan, Sudan and Egypt [4].

Much of the literature on water resources in Uganda is scattered and embedded in policy documents, strategic plans and reports produced by consultants and international organisations. There are also several issues and challenges surrounding water resources, which are aggravated by climate change and variability and population growth among others. The current water management practices in Uganda may not be robust to cope with these challenges which impact on water resources and increase water use requirements. With rapid growing population and improving living standards, the pressure on Uganda's water resources is increasing and per capita availability of water resources is reducing day by day. Spatial and temporal variability in precipitation is another challenge the country faces, which often result into floods, landslides and droughts. The quality of surface and ground water resources is also deteriorating because of increasing pollutant loads and non-point sources. Climate change is expected to impact on rainfall and water availability. Currently, the data collection, processing, storage and dissemination have not received the deserved attention. All these aspects require an assessment of the current water resources status. The review therefore, focuses on the character, potential, governance, challenges, availability and demands of water resources of the country using secondary data sources.

2. Hydrology of Uganda

2.1. Surface Water Resources

Uganda is a landlocked country that occupies 241550.7 km^2 of land. Open water and swamps constitute 41743.2 km^2 of area [4], with about 16% of total land area of wetlands and open water, plus the annual water supply of 66 km^3 in form of rain and inflows. One would therefore conclude that, Uganda is fairly well endowed with water resources. Since direct rainfall is the most important source of water resources in Uganda, understanding the

Table 1. The major shared water bodies and river courses of Uganda.

Water resource	Countries sharing	Basin wide
River Nile	Uganda, South Sudan, Sudan, Egypt	Burundi, DRC, Egypt, Ethiopia, Kenya, Rwanda, South Sudan, Sudan, Tanzania, Uganda, Eritrea
River Achwa	Uganda, South Sudan	Uganda, South Sudan
River Kagera	Tanzania, Burundi Rwanda, Uganda	Nile Basin Riparian States-Burundi, Tanzania, Rwanda, Uganda
River Semliki	DRC, Uganda	Nile Basin Riparian States
River Malaba	Kenya, Uganda	Kenya, Uganda
River Sio	Kenya, Uganda	Kenya, Uganda
Lake Victoria	Uganda, Kenya, Tanzania	Nile Basin Riparian States
Lake Albert	DRC, Uganda	Nile Basin Riparian States
Lake Edward	DRC, Uganda	DRC, Uganda
Sango Bay Swamp	Tanzania, Uganda.	Kenya, Burundi, Rwanda

spatial and temporal variability of rainfall is therefore paramount when assessing water availability [5]. Water availability has determined the local water resources, land-use potential and population distribution.

The Department of Water Resources Management of Uganda has categorized the surface water resources into eight main drainage sub-basins [4]. These include; Lake Victoria, Lake Kyoga, River Kafu, Lake Edward, Lake Albert, River Aswa, Albert Nile and Kidepo Valley (**Figure 1**). The yield from these sub-basins, though small compared with the total Nile flow, dominates the water resources potential with in Uganda. Major water bodies include lakes Victoria, Kyoga, Albert, George, Edward (**Table 2**) and another 149 smaller lakes spread across the country covering an area of 38,500 km^2 [3]. The lakes are interconnected by a river system, which developed as a result of river reversal and ponding that formed the two major lakes [6] (and references there in). In the north-eastern part of the country, many of the water courses are seasonal. The main river basins of Uganda are

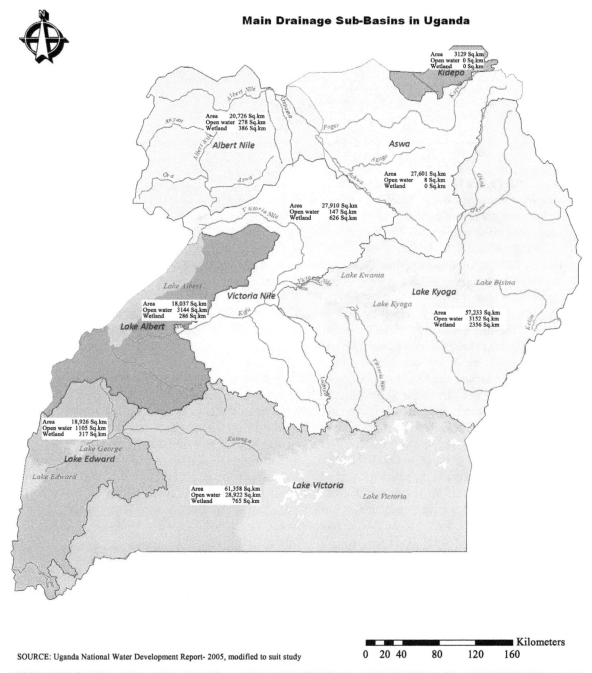

Main Drainage Sub-Basins in Uganda

	Kidepo
Area	3129 Sq.km
Open water	0 Sq.km
Wetland	0 Sq.km

	Albert Nile
Area	20,726 Sq.km
Open water	278 Sq.km
Wetland	386 Sq.km

	Aswa
Area	27,601 Sq.km
Open water	8 Sq.km
Wetland	0 Sq.km

Area	27,910 Sq.km
Open water	147 Sq.km
Wetland	626 Sq.km

	Lake Kyoga
Area	57,233 Sq.km
Open water	3152 Sq.km
Wetland	2356 Sq.km

	Lake Albert
Area	18,037 Sq.km
Open water	3144 Sq.km
Wetland	286 Sq.km

	Lake Edward
Area	18,926 Sq.km
Open water	1105 Sq.km
Wetland	317 Sq.km

	Lake Victoria
Area	61,358 Sq.km
Open water	28,922 Sq.km
Wetland	765 Sq.km

SOURCE: Uganda National Water Development Report- 2005, modified to suit study

Kilometers
0   20 40   80   120   160

Figure 1. Main drainage sub-basins in Uganda including rivers and lakes.

Table 2. Major Lakes of Uganda and their associated characteristics.

	Surface Area (km²)	Area in Uganda (km²)	Mean Elevation above Sea Level (m)	Maximum Depth (m)	Volume in Uganda (km³)	Mean Depth (m)
Lake Victoria	68,800	28,655	1134	84	1237	40
Lake Albert	5659	2850	618	56	80	25
Lake Kyoga	2636	2636	1034	10.7	7.9	3
Lake Edward	2324	638	912	117	16.8	34
Lake Kwania	540	540	1033	5.4	2	4
Lake Wamala	250	250	1290	9	1.2	4
Lake Bisina	150	150	1030	N/A	0.5	N/A
Lake George	228	228	914	7	0.8	2.4
Lake Bunyonyi	61	61	1974	39.3	0.2	
Lake Kachira	39.6	39.6	1235	4.8	0.2	
Other 149 Minor Lakes		2453			7.2	
Total		38500.6			1353.8	

Source: UN-Water, 2006 in [6].

shown in the **Figure 1**.

The most prominent hydrological feature in Uganda is Lake Victoria, which is the second largest fresh water lake in the world [7], with an area of 69,000 km². River Nile, which is the only outflow from the lake, has its source at the point where Lake Victoria spills over Ripon Falls. The 130 km stretch of the Nile from Lake Victoria to Lake Kyoga is termed the Victoria Nile. Lake Kyoga is drained through the Kyoga Nile which, after a relatively flat reach downstream from the lake, enters a series of rapids and falls before it flows into Lake Albert at a level 410 m lower than Lake Kyoga. The plateau immediately to the north of the Rwenzori drains to Lake Albert via the Muzizi River.

In Lake Albert, the Nile is joined by river Semliki which drains Lakes George & Edward found in the rift valley and the high rainfall area of the Rwenzori Mountains. Lakes George & Edward are connected through the Kazinga Channel. The Nile flows from Lake Albert with a gentle slope to the Sudanese boarder. This reach of the river is called the Albert Nile.

Major rivers include the Nile, Ruizi, Katonga, Kafu, Mpologoma and Aswa. The rivers and lakes of Uganda are within the upper part of the White Nile Basin [4], with the exception of those in the tiny North-eastern catchment of Kidepo, which drains into the Lake Turkana basin in Kenya (**Figure 1**). The North-western slopes of Rwenzori watershed in Uganda drain to Lake Edward via the Ishasha, Chiruruma, Nchwera and Nyamweru rivers, and also by several streams, which enter the western flowing part of the Katonga River. The North-eastern part of the Virunga range watershed however, drains directly to Lake Victoria via a series of swampy lakes and streams culminating in the Kibaler River, which enters Lake Victoria through the swamps at Sango Bay [8]. The Ugandan slopes of Mt. Elgon and the central highlands along the Kenyan border drain via rivers with swampy valleys or seasonal floodplains to Lake Kyoga, while the north eastern highlands and most of the northern plateau drain directly to the Bahr el Jebel in Sudan via the Achwa River that also joins the Nile.

Because of warping of the landscape, many of the perennial streams of the plateau are clogged with swamp [6]. About 10 Percent of the country is covered by swamps (wetlands), of which one third is permanently inundated. In the south and west of the country, swamps form an extensive low gradient drainage system in steep V-shaped valley bottoms with a permanent wetland core and relatively narrow seasonal wetland edges. In the north, they mainly consist of broad flood plains. In the east they exist as a network of small, vegetated valley bottoms in a slightly undulating landscape [6].

2.2. Wetland Resources

There are basically two broad distributions of wetland ecosystems in Uganda: 1) the natural lakes and lacustrine swamps found around major lakes; 2) the riverine and flood plain wetlands which are associated with the major river systems in Uganda. Wetlands, cover about, 29,000 km^2, or 13% of the total area of the country [9]. The wetlands comprise swamp (8832 km^2), swamp forest (365 km^2) and sites with impeded drainage (20,392 km^2). Wetlands also include areas of seasonally flooded grassland, swamp forest, permanently flooded papyrus, grass swamp and upland bog. By 2008, the coverage had declined to 10.9% (**Table 3**). This decline has been generally observed around Lake Victoria and Kyoga drainage basins. The decline according to the Uganda water and environment sector performance report of 2011 is largely attributed to encroachment for expansion of urban centres, settlement, industrial developments and extension of agricultural land. Except for Sango Bay, the bulk of Uganda's wetlands lie outside protected areas [3].

Wetlands perform a number of functions, such as; mitigating the effects of both floods and droughts, providing fish resources and support cropping and grazing along their margins. They are centres of high biodiversity and productivity as well as valuable refuges and sources of food like fish. Furthermore they are active biological filters in the treatment of effluents, but due to this function they are also sensitive to the accumulation of pollution.

2.3. Yield of Rivers

All rivers in Uganda ultimately reach the Nile through various interconnections. Thus an understanding of the Nile flows, gives us an insight into the flow regime of major rivers in Uganda. The flows of the Nile are highly variable from year to year. This annual variability has been monitored and reported in previous studies [3] [4] evidenced in **Figure 2**. The plot shows the long-term variations in flows monitored along Victoria Nile at Jinja, Kyoga Nile at Masindi Port and Albert Nile at Panyango. It shows the flow increased from 1960s, after which there is a steady decline in the flow at all monitoring points. The long-term average outflow from Lake Victoria has been 840 m^3/s and the range of outflows is between a minimum of 345 m^3/s and a maximum of 1720 m^3/s. At the 95% monthly reliability level, the flow of the Victoria Nile is of the order of 495 m^3/s.

Using annual rainfall data series for the period 1940 to 1999, we computed a Drought Severity Index (DSI) for the drainage sub-basins in Uganda. Results presented in **Table 4** have been compared with plot in **Figure 2**, which was produced by DWRM. There is a close correlation between DSI and the river Nile outflows monitored for 1899-1997 period.

Table 3. Changes in wetland coverage by drainage basin between 1994 and 2008.

Drainage Basin	1994 (Area km^2)	%	2008 (Area km^2)	%	% Change
Albert Nile	1736.3	6.21	1255.2	4.71	27.7
Aswa	3028.0	10.83	2168.9	8.24	28.4
Kidepo	168.1	0.60	197.2	0.74	−17.3
Lake Albert	2838.6	10.15	2421.7	9.20	14.6
Lake Edward	1671.1	5.97	1096.3	4.16	34.4
Lake Kyoga	15008.3	53.67	11028.5	41.92	26.5
Lake Victoria	7167.6	25.63	3310.2	12.58	53.8
Victoria Nile	5786.3	20.69	4829.4	18.35	16.5

Source: Uganda water and environment sector performance report, 2011—modified to suit study.

Table 4. Drought years for selected drainage sub-basins in Uganda.

Series Variables	Climatic Normal	Albert Nile	Lake Kyoga	Lake Victoria
Annual	1940-1969	1943, 1949, 1952, 1953, 1954, 1965	1943, 1949, 1953, 1957, 1958, 1965	1943, 1946, 1952, 1953, 1959, 1965
	1970-1999	1971, 1983, 1984, 1989, 1993	1973, 1974, 1979, 1980, 1984, 1986, 1993	1970, 1971, 1980, 1982, 1983, 1985

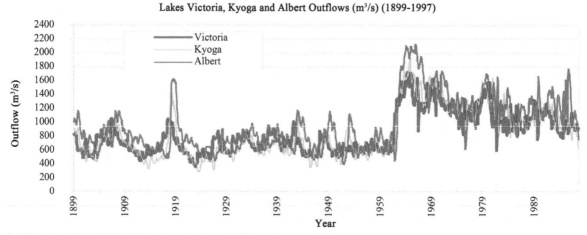

Figure 2. Long-term variations in river Nile flows (figure adopted from WRM subsector reform study report 2004).

Following steps in [10], one can discover the years and seasons when the sub-basins were experiencing droughts. It is evident from the plot (**Figure 2**) and **Table 4** that, the early 1940s and 1950s in the first climatic normal were years of below normal rainfall. The situation is similar in the second normal when the early 1980s were drought years, consequently affecting the outflow of river Nile. This re-affirms the fact that the main source of water for the water resources in Uganda is rainfall and the whole of Uganda's water resource is part of the Nile.

Annual rainfall data for the period 1940 to 1999 were exposed to Sen's T test to establish the behavior of rainfall trends in Uganda. The results of Sen's T test applications to annual rainfall of the main drainage sub-basins are shown in **Table 5**. The Sen's T test results indicate a negative trend of annual rainfall in six of the sub-basins, namely; Albert Nile, River Aswa, Lake Albert, Lake Edward & George, Lake Kyoga and Lake Victoria. Significant decreasing trends were observed in River Aswa and Lake Albert sub-basins. Results for Victoria Nile sub-basin show a positive trend, which is not statistically significant at 0.05. Consequently the decreasing rainfall affects water resource distribution in the country.

Despite Uganda's significant water resources, their spatial and temporal variability often renders many parts of the country water stressed over long periods of the year. Water resource management sub-sector reform study of 2004 showed that, districts in the north-eastern and south-western parts of the country have the least per capita water availability. The study also reveals that by 2015 more than 75% of the country will be water stressed (**Figure 3**). The methodology used in computing the water availability only considers the runoff generated within each district. It is used here to give a picture of the average yield of major catchments in Uganda.

For example the south-western and north-eastern parts of the country have the lowest annual run-off (<1 litre/s/km^2). These are semi-arid areas, which receive very low rainfall. The Lake Victoria basin has the highest run-off (>10 litres/s/km^2) [11]. Mean annual rainfall calculated for the different drainage sub-basins during 1940 to 2009 is depicted in **Figure 4**. Analysis supports the correlation between rainfall and run-off generated from each sub-basin. The average yield of major catchments in Uganda is also represented in **Figure 4**.

A recent study by DWRM in Rwizi, Wamala and Victoria catchments, in the Victoria water management zone concluded that, reduced water levels occur annually in the dry season with in the three catchments [12]. The lowest levels are witnessed at the end of every dry season, until the rains set in, but not reaching the original high levels. This however paints a bleak picture especially when the rainy seasons become unpredictable [12].

3. Status of Water Availability

Two organizations, Directorate of Water Development (DWD) and National Water and Sewerage Corporation (NWSC) have been tasked to ensure that water is availed to Ugandans. Available information from NWSC shows that there has been a steady increase in total water production e.g. from 46.7 million·m^3/yr, in 2000/01 to 72.14 million·m^3/yr in 2009/10 financial years. Comparing the water requirements with the available renewable freshwater, Uganda had a capacity to utilize only 1% of the current renewable freshwater for consumptive pur-

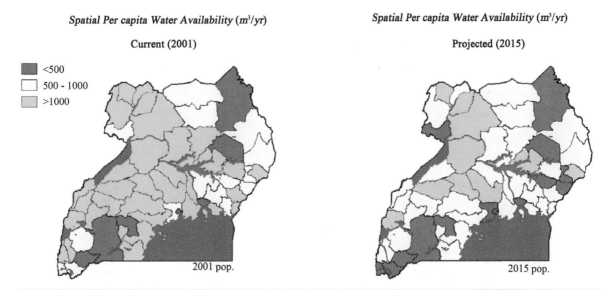

Figure 3. Spatial per capita surface water distributions (m³/yr). Source: WRM sub-sector Reform study report, 2004.

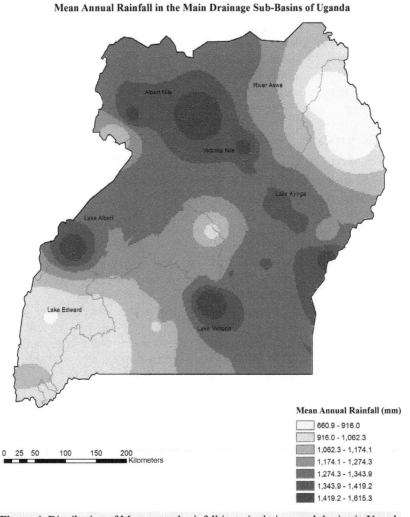

Figure 4. Distribution of Mean annual rainfall in main drainage sub-basins in Uganda 1940-2009.

Table 5. Trend results for the main drainage sub-basins. Sen's T-test was performed on area average rainfall data.

Sub-Basins	Rain Series	Albert Nile	River Aswa	Lake Albert	Lake Edward	Lake Kyoga	Lake Victoria	Victoria Nile
Year of Rainfall	Maximum	1961	1983	1961	1996	1961	1961	1961
	Minimum	1943	1943	1943	1979	1957	1957	1943
SEN's T-Test	Area mean	−0.31	−3.91	−3.13	−0.91	−1.09	−1.36	0.19

poses. Out of the total water withdrawal, domestic water supply accounted for about 51%, agriculture 41% and industry 8% [8]. Withdraw per capita rose from 12 m^3/yr in 2002 to 21 m^3/yr in 2008 (**Figure 5**).

Percent change is equivalent to 112% in a seven-year period. Notable changes were realized in domestic withdrawals from 45% to 51%, while industrial withdrawals fell by 7%. The use of water all over the world is a function of population change, food consumption, economic policy, technology, lifestyle and societies' views of the value of freshwater. Since Uganda has one of the fast growing populations (3.2%), its demands will lead to increased water withdrawal. For instance, total volume of water sold according to NWSC annual reports had increased from 31,151,380 million·m^3/yr in 2002/03 to 47,027,817 million·m^3/yr by June 2010. Much of the total water produced was sold to domestic consumers (48.9%), industrial and commercial enterprises (25.4%), institutions and governments (20.8%), and (4.8%) was sold through public standpipes. This positive trend shows how effective the organizations have been in achieving national targets.

The major water sources exploited are protected springs, deep boreholes and shallow wells. These are also sources through which natives access water as individuals or licensed independent water providers. Records from water permit database of the DWRM, for example put water abstractions of registered users for the year 2003 in Lake Albert basin at 1500 m^3/yr abstracted from surface water while ground water abstraction accounts for 63 m^3/yr. Water abstraction particularly from the Albert basin is expected to increase following the recent discovery of oil reserves. Water resources management actions within the Albert basin will require vigilance and strict monitoring during the process of drilling and processing oil.

3.1. Groundwater Occurrence

The primary source of freshwater for drinking and irrigation in the world is groundwater. Ground water supplies 75% of all safe sources of drinking water in Africa [13]. In Uganda for example, 61% of the country's water is from a ground water source, accessed from springs and boreholes around Lake Victoria and south-western Uganda. Several studies, like [13]-[17] have assessed ground water occurrence in Uganda from different perspectives mostly at a catchment scale. These assessment studies have been a basis for water resources planning in the country.

According to the Monitoring and Assessment Division report of 2011, ground water resources of Uganda were estimated in 2010 during the National water resources study to be as shown in **Table 6** for the different basins. The report produced a map of estimated exploitable groundwater resource per district. This exploitable ground water represents the proportion of renewable resources that can be exploited on a sustainable basis without seriously affecting the environment.

The Victoria Nile and Kyoga basin have sustainable ground water, which is more than 36 mm/yr, while Kidepo has the least amount of sustainable ground water equivalent to 6.3 mm/yr (**Table 6**). This is a true reflection of the catchment size from which these basins draw water (**Figure 1**).

3.2. Groundwater Development

Groundwater is the major source of water supply in the rural, semi-arid and arid areas in Uganda. Groundwater development has been going on since the 1930s through construction of deep boreholes, shallow wells and protected springs. There are approximately 20,000 deep boreholes, 3000 shallow wells and 12,000 protected springs in the country constructed mainly for rural domestic water supply. In the 1990s ground water was intensified to provide for town water supply. For example, 782 small towns were identified for the provision of piped water by June 2006, and 70% of this water was to be provided by ground water sources like the deep boreholes [13]. Boreholes and shallow wells are normally installed with hand-pumps with capacity of 1 m^3/hour and their yields

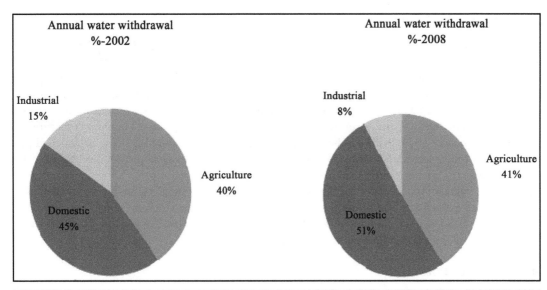

Figure 5. Water withdrawals in Uganda. (Data was sourced from WRMD quarterly reports, joint water sanitation sector program support, DWRM).

Table 6. Average sustainable groundwater for major river and lake basins.

Drainage Basin	Area km^2	Land Area km^2	Sustainable Ground Water (mm/yr)
Lake Edward	18,946	17,855	20.3
Lake Victoria	61,886	32,924	24.7
Lake Albert	18,079	14,882	23.7
Victoria Nile	27,961	27,807	39.9
Lake Kyoga	57,236	53,899	36.1
Albert Nile	20,727	20,484	24.4
Aswa	27,637	27,635	17.3
Kidepo	3229	3228	6.3
Miscellaneous	5716	5679	15.0

Source: Monitoring and Assessment Department, DWRM.

are usually low. Ground water has advantages like, good quality water that requires little or no treatment costs can easily be operated and managed by users, which makes investment and operational costs to be lower than those of surface based systems [13]. Boreholes with yields greater than 3 m^3/hour are thus normally considered for installation with motorized pumps for piped water supply.

In 1996 GoU initiated groundwater assessment studies, to fully understand the nature, extent and reliability of the country's groundwater resources. The study has provided Information on the distribution and behaviour of aquifers, groundwater recharge, aquifer vulnerability to pollution, impact of motorized abstraction on groundwater resources and a conceptual model of groundwater dynamics has been developed.

For example, groundwater assessments in three catchments namely Ruizi, Wamala and Victoria concluded that, groundwater resources were declining [11]. The causes of decline according to the monitoring and assessment department of the DWRM report, includes among others; climate change/global warming, changes in land-use, especially deforestation, unsustainable water withdrawals, poor catchment management, prolonged droughts, reduced rainfall in the catchment (see **Table 5**). This information, though still scanty, forms the basis for the current groundwater resources planning and management in the country.

3.3. Groundwater Recharge

Recharge assessments have recently been carried out in Apac in northern Uganda, Mbarara in western Uganda, Wobulenzi in central Uganda, Nkokonjeru in eastern Uganda and Hoima in mid-western Uganda. Results obtained using the various methods on groundwater recharge reveal a range between 90 and 220 mm per annum and accounts for 7% and 20% of the average annual precipitation in Uganda [18]. Dynamically downscaled climatological data have been applied on a soil moisture balanced model to estimate ground water recharge and runoff in south-western Uganda along river Mitano [13]. Their results under the A2 emissions scenario, for the period 2070-2100, relative to 1960-1990 indicated an increase in precipitation (14%) and a modelled potential evaporation (53%) with a substantial increase in rain intensity [13]. Hydrological projections also gave a rise to increases in recharge (53%) and runoff (137%).

From the above figures it can be stated that groundwater recharge in Uganda is quite high compared to current abstraction volumes and will not be a limiting factor in groundwater development for a few years to come. However, there is a need to carry out more detailed recharge and water balance studies in the country to ensure that groundwater development is carried out in a sustainable manner.

3.4. Ground Water Potential

The potential of groundwater in various areas of the country is exhibited by presence of deep boreholes, shallow wells and springs. These are assessed in the paragraphs that follow.

3.4.1. Deep Boreholes

Deep borehole potential can be assessed by a means of a number of borehole parameters such as regolith thickness, aquifer yields and rest water levels. Uganda is characterised with a clayey regolith especially in the upper layers where relatively low permeability dominates. The regolith thickness across the country can be described as low to medium varying between 20 - 45 m. This leads to medium to high ground water potential through provisional storage. Aquifer yields on the other hand vary from one part of the country to another according to the formation in which they are drilled and their degree of fracturing and weathering. For example, potential yield from deep aquifers were estimated during the AQUASTAT survey, to be above 3 m^3/hr in the southwest, southeast, northwest and along the eastern border of the country [8]. In large areas of central parts of the country potential yields are between 2 and 3 m^3/hr, while in some areas it is below 1 m^3/hr [8]. Rest water levels also give an indication of the groundwater potential of an area. Shallow water levels (<20 m) indicate that the aquifer has high potential for yielding groundwater while deeper water levels indicate the reverse. Rest water levels (static water levels) in the country vary between 1 and 45 m below ground level.

3.4.2. Springs

Springs occur either where the flow of unconfined groundwater is interrupted by an impermeable formation or where the head of confined groundwater is released by flow to the surface [8]. There are 2 major types of springs in Uganda namely; contact and fracture springs. Fracture springs are usually very susceptible to contamination and drying up while contact springs are more reliable [14] [16].

3.4.3. Shallow Wells

The potential of shallow wells is quite high, especially in the valleys. Their potential is favoured by the thick regolith that is fairly coarse grained. From Uganda's experience, shallow wells are a very reliable source of water supply to the communities although precautions need to be taken to ensure that they are not contaminated. Springs and wells are a major source of water for the country [4]. Through DWD, shallow wells have been constructed and improved to provide good quality water.

4. Water Resources Sector Development Trends

Much of the national documentation in Uganda refers to a series of development plans, strategies and action plans which were planned and executed to varying degrees of success, without an explicit sector policy framework before the 1980s. In the 1990s a number of guidelines were prepared and studies undertaken by government which paved the way for the review of water legislation and sector policy development process. Since then,

GoU has taken major steps to rationalize water resources management, development and the delivery of Water and Sanitation Services (WSS). The development is aligned to achieve the Millennium Development Goal (MDG) of 62% by 2015 on water.

To achieve the set targets, the GoU has reformed the water and sanitation sector in order to ensure that services are provided and managed with improved performance and cost effectiveness. In this regard reform studies were carried out that resulted in development of Sector Strategic Investment Plans (SIPs) with appropriate policies and strategies like;

- The rural water and sanitation SIPs which aim at increasing the safe water supply coverage to 77% or 95% by 2015.
- The urban water and sanitation reform SIPs that aim at providing 100% urban population with safe water supply and sanitation by 2015.
- The water for production SIPs, which stipulates the participation of the private sector and farming community in financing this regard.
- The water resources management SIPs, which aims at promoting integrated water resources management and development at both national and local levels.

The strategies above required a Sector Wide Approach to Planning (SWAP). SWAP framework for the WSS sector was adopted at the second GoU/Donor review for the water sector in September 2002. The move towards sector wide approach called for GoU/donors to promote; uniform fund disbursement rules, uniform and stronger fund accountability rules, common indicators, joint appraisals and reviews. Since then reviews have been carried out every year to monitor progress in the water sector.

The National Water Policy was adopted in 1999 and sets the stage for water resources management and guides development efforts aimed at improving water supply and sanitation in Uganda. It promotes a new integrated approach to water management that guides the allocation of water and the associated investments. This new approach is based on the continuing recognition of the social value of water, while at the same time giving much more attention to the economic value of water. Key aspects include; application of a participatory demand driven approach to planning which promotes user ownership and management of services, backed by measures to strengthen local authorities and private institutions in implementing and sustaining water and sanitation programmes.

The National Water Policy embraces international and regional resolutions, declarations and guidelines [19] because more than 90% of Uganda's waters are trans-boundary or river courses. This therefore means that the use and development of Uganda's water resources is controlled by international laws and principles and colonial agreements.

Because much of Uganda's water ends up with the Nile, therefore, Uganda's approach to the water resources development and management is a critical element within the Nile Basin and hence requires clear trans-boundary policy and institutional frameworks and also legal and harmonized laws with riparian states.

The current strategic framework for the development and management of water resources in Uganda recognizes two important principles. First, water is fundamental, to achieving the national objective of poverty eradication through the promotion of rapid economic growth, good health, food security, and social equity. Secondly, within the regional context, according to the ministerial policy statement, cooperative development of the trans-boundary water resources e.g. the Nile, the Kagera, Sio-Malaba River, Lake Victoria, Lake Albert, Lake Edward and others can serve as a catalyst for a broader range of cooperation and economic integration.

5. Water Resources Utilization in Uganda

Water for Agriculture/irrigation: Rain fed agriculture is the most practiced land use method in the Uganda. Currently this practice is threatened with climatic variability and a fast growing population, which are impacting on food security levels. As a result national policy on agriculture aims at increasing agricultural production per unit area e.g through a more efficient use of land and water resources which will improve food security. The increase in population has increased the need to raise crops in areas that do not get enough rainfall, hence requiring irrigation. A Food and Agricultural report study in 1987, estimated the irrigation potential in Uganda to be more than 280,000 ha. If fully developed a water demand of 2.7 billion·m³/year will be required from Uganda's water resources hence, placing a considerable demand on the national and trans-boundary water resource like the river Nile. In order to manage water fairly, the Shared Vision Program (SVP) on efficient water use for agricultural production was set up to establish a forum to assist stakeholders at regional, national and community levels to address issues related to efficient use of water for agricultural production in the Nile basin and link it with the trans-boundary water policy. This program provides the appropriate linkage under the Trans boundary Water Resources Program (TBWRP) being formulated with NBI-SVP for improved agricultural Production.

Water for power: Most waterpower is used to generate Hydro-Electric Power (HEP), which is used to light homes and to run factories. A study by Nile Basin Capacity Building Network (NBCBN) in 2005 estimated the hydrological resources to have a power production potential of over 2500 MW, of which over 2000 MW is mainly concentrated on river Nile and the rest is scattered in parts of the country.

By 2005, less than 10% of this potential was exploited [20]. Of the 2000 MW potential along the river Nile, 630 MW (31.5%) have been tapped, and the unexploited potential is well over 1300 MW [21]. Installed capacity stands at 872 MW including thermal and hydropower plants, of which HEP contributes 621.5 MW. This shows that power supply of Uganda is almost totally dependent on hydropower. Unfortunately, installed capacity is grossly inadequate to meet the current and rapidly increasing demands. For example in 2008, the number of consumers by type was recorded at 347,433 for domestic small general service; 28,810 for commercial small general service; 324 for large industrial; 1194 for general-medium industrial and 195 for street Lighting in 2008 [22]. Given the current economic growth of 7%, domestic power demand is expected to increase to 1130 MW in 2023 [23]. One of the national development plan objectives is to increase power generation to between 780 MW and 820 MW, as well as increase rural electrification by 10% and promote efficiency programmes and renewable sources countrywide.

The main sites that have been identified for the development of major schemes are all located downstream along the river Nile. The other small-scale hydropower potentials exist along the rivers draining Mt. Elgon, the extreme southwest of Uganda, rivers draining west Nile near Arua and the rivers draining the Rwenzori Mountains. The country has a big potential for medium, small and micro hydropower stations. Development of large schemes on the Nile should be the right way to go by the government of Uganda, because the markets are available in the region.

The national policy provides for regional co-operation for optimum hydropower development. For example, the Nile Basin Power Trade of NBI-SVP has the objectives of establishing the institutional means to coordinate the development of regional power markets and improve access to reliable and low cost power among the Nile Basin countries. In order for Uganda to fully exploit her hydropower potential on the Nile, this program will be the main avenue for the linkage for cooperation under the trans-boundary water policy and to also promote the benefit-sharing concept.

Water for industry: The role of water for example in industry contributed 24.2% in 2008/2009 to the total Gross Domestic Product (GDP) and increased to 26.9% in 2010/2011 [24]. Water is highly consumed in the construction industry [25] especially during this period when the country is undergoing infrastructural development. These industrial activities (Construction, manufacturing, mining and quarrying) that consume a lot of water contributed 12.3%, 7.5%, and 0.3% of the total GDP in 2008/09 respectively. Water, which is vital in transforming industry and eradicating poverty, will be in high demand in future, given the trend of economic growth.

Water for fish: Inland water bodies support a thriving fishing industry in Uganda. Fish catches vary from place to place, with the largest yields associated with major lakes and rivers (**Table 3**). According to [24] fishing activity was worth 2.9% share of the GDP by 2008/09 fiscal year and many people are indirectly dependent on this industry. Fish is often a large part of diet of people living in the country and abroad. Tonnes of fish are also harvested from wetlands and ponds on fish farms. Fishing activities have become so popular that are fast increasing in importance and numbers that exerts pressure on water resources especially on quality of water resources.

Water for transport: People depend on water transport to carry goods for use and trade from one region to another, using boats, ferries and ship. Water transport links Ugandans living on islands with the mainland and also those involved in trade across borders towards DRC, Tanzania and Kenya. Transport as an activity though not fully developed, has impact on the quality of water resources, which requires further study.

Water for environment and tourism: Development of tourism in Uganda is a high priority area. Tourism development is dependent on the natural beauty and quality of the environment, including wetlands, range lands, lakes, rivers and the associated sceneries like the waterfalls, beaches and rare fauna species which provide an array of tourism activities. The splash and flow of water in the streams, hot springs, geysers and fountains also soothes and inspires many people who love being near water. Most of the recreational places in Uganda provide water sports, such as swimming, fishing, Water rafting and sailing. People also enjoy the beauty of a quiet lake, a thundering waterfall (like at Karuma, Kabalega and Budhagali) or even roaring surf. Business community in Uganda has built recreational areas along the lakes, such as Entebbe, Lutembe, Sesse-Kalangala, Kigo, Jinja where local tourism is flourishing. This study discovered that climate variability was impacting on the activities at the shores of the Lake Victoria, and the subsequent income of the entrepreneurs.

These touristic facilities generate huge financial and socio-economic benefits for the country. For example in the year 2008, a total of 844,000 tourists to the country were recorded, of which 144,000 declared their purpose of visit as leisure and holiday. And about 138,000 persons (foreign and national) visited the national parks in 2008, 190,000 in 2010 and 208,000 in 2011. Between 2010 and 2011 the industry grew by 9.4% and contributed 21% to GDP in 2011, earning the country US $805 million in 2011/12 financial year according to the Daily Monitor Newspaper, 26.09.2012.

Water for Sanitation: This is a field of public health necessary for controlling and preventing disease. It includes personal cleanliness, sewerage treatments, waste disposal systems, water treatment and so forth. Sanitation deteriorates where water conditions are poor and uncertain which directly affects the wellbeing of communities. Likewise sewerage and sanitation service requirements increase in step with improvements in water supplies and have important health implications. Because water, health and sanitation are set targets for 2015 MDGs, African Development Bank (ADB) financed water and sanitation projects to small towns and rural areas in Uganda to the tune of USD 28 million and USD 60 million respectively [26]. GoU has also signed agreements in 2012 to finance projects to improve health services delivery and access to water and sanitation in both urban and rural areas worth USD 155.8 million of investment [27]. This development is expected to put more pressure on the water resources of the country as discussed in sections above.

Water supply: Water supply is a key factor of production in e.g. manufacturing industry, power generation, mining, construction and agriculture. It sustains the natural environment; therefore its quantity and quality are important indicators of what the country has achieved. Access to water has levelled off at 64% for rural and 77% for urban areas but set targets are 100% and 77% for urban and rural areas respectively by 2015.

The challenges faced in Uganda with regard to water supply issues are: to ensure that water for both human and livestock consumption is available in sufficient quantities and quality all the time; that the demand is quantified; water is used efficiently and sanitation is promoted; and shared waters and watercourses are used and developed to meet this demand in particular with regard to the new strategy of bulk water supply.

6. Vulnerability to Climate Change and Adaptation Strategies

The changing climate in Uganda presents very serious national challenges and risks across various sectors such as agriculture, water resources and energy, which support the economy and the wellbeing of its people. Climate change has significant impacts, which manifest itself in form of droughts, variations in groundwater and surface water levels, melting of ice caps and incidences of diseases, high temperatures and wild fires. For example, [10] identified 8 seasonal droughts within the Lake Victoria basin in the period between 1990 and 1999. Our analysis has also established drought years and seasons in the various drainage sub-basins in Uganda (**Table 4**). Prominent was the 2004/5 drought period when the water level in Lake Victoria dropped by a meter below the 10 year average [28]. Droughts significantly affect the country's water resources, hydropower production and agriculture.

Cycles are observed in rainfall patterns in Uganda for the period 1940-2009. Analysis shows decades of normal and beyond normal rainfall (**Figure 6**) during the period of analysis. Three long spans of below normal rain are evident in the 1940s, 1960/70 and 1980s. Other particularly wet periods were experienced in early 1950s, 1960s, late 1970s and the 1990s (**Figure 6**). The contributing factors to the below average annual total rainfall for Uganda have not been established in this study despite the 1980s being drought years with consequences on food security for the continent [29].

From a hydrological perspective, Lake Victoria, the largest water resource exerts a big influence in Uganda's climate. The main source of water for the lake is rain, but due to rainfall anomalies, the lake has consequently displayed large and rapid changes. This has been demonstrated by [30] [31] who find a significant correlation between lake rainfall series and lake levels. A number of studies highlighted by [15] and [32] have concluded that the lake is a sensitive indicator of climate change. Models on Sensitivity of lake levels, predict a fall in le-

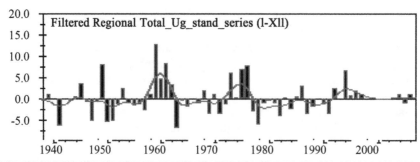

Figure 6. Regional total rainfall (mm) for the period 1940-2009.

vels during the 2021-2050, and levels pick up by 2070-2099, time horizons in both emission scenario. This circumstance is explained by the predicted decrease in direct rainfall with a corresponding increase in lake potential evaporation by 2021-2050 [15].

Melting glaciers have been a centre of attraction of late as a source of water. The interaction between glaciers and climate represents a particular sensitive process on the water resources of Uganda. A number of authors have been cited by [11] to have comprehensively studied the effect of this interaction in the Rwenzori catchment area. Recent research in the 1950s, 1990s and 2000s indicate that the area covered by alpine glacier has reduced from 7.5 km^2 in 1906 to 1 km^2 in 2003 on Rwenzori Mountains [33]. Between 1955 and 1990, Rwenzori ice caps according to the Sector Performance Report of 2011 have retreated by 40%. This drastic trend calls for further studies and urgent response from policy makers.

The effect of deglaciation on water resources has been investigated to determine the impact of glacial melt on river discharge during the dry season and the risk of flooding during the rainy season [34]. Their findings, also held by [35] showed that current glacial recession has a negligible impact on alpine river flow since melt water discharges contribute less than 2% of the total discharge of the principal river receiving melt water (River Mubuku) at the base of Rwenzori Mountains during both dry and wet seasons. Consequently the effect of melt water has insignificant impact on the water resources, but may have adverse effect on tourism of Uganda. Melting of ice caps of the Rwenzori is expected to have an effect on water catchments downstream and eco-tourism as well as the overall economy [36].

The effects of climate change on ground water recharge and base flow in upper Ssezibwa catchment in Uganda has been studied using global climate models [17]. Projections from their study indicate that rainfall will generally increase, from 20% to 100% by the 2080s. Simulations from the same study showed an increase in recharge, base flow and total flow in the future. The mean annual daily discharge is expected to increase by 40% - 100% from the current 1.47 m^3/s in the coming 20 to 80 years.

7. Water Governance in Uganda.

Water governance is seen as a range of political, social, economic and administrative systems that are in place to develop and manage water resources and the delivery of water services at different levels of society [37]. The manner of allocation and regulatory process and interaction in the management of water resources that embraces formal and informal institutions [38] including poverty eradication [39] are a focus of governance. Some of the necessary conditions for good governance are inclusiveness, accountability, participation, transparency, predictability and responsiveness. A major challenge is to understand how all these different processes in concert, determine certain policy outcomes and how change in governance regimes occurs.

For instance policies tend to focus on standard policy solution like liberalization and pricing of water services, which have had counter-productive results. Currently Climate change and the related increase in extreme weather events have exposed the vulnerability and lack of resilience of water resource management regimes. At the same time, policies to accommodate the prospects of climate and global change are absent. Consequently the conditions under which water resources managers have to perform have become increasingly unpredictable especially in a developing country like Uganda.

The broad strategies in Uganda for implementing activities and programs in the Water sector are articulated in the Poverty Eradication Action Plan (PEAP) of 2004. PEAP which formed the basic planning framework for the formulation and implementation of all sector policies, strategies and programs, identifies water and sanitation as part of the priority areas for poverty eradication. The strategies are embedded in the three pillars, namely; to increase production and competitiveness of Uganda's products to attain increased household incomes including water for production and water resources management; Strengthening security, conflict resolution and disaster management (includes water for security in north-eastern Uganda and provision of water and sanitation services; and human development among others.

Implementation of the PEAP is guided by a comprehensive sector policy and legal framework for management of the water and sanitation sector. These include: the 1995 Constitution of the Republic of Uganda, Local Governments Act Cap 243, Water Act Cap 152, and accompanying regulations, the 1998 Water Resources Regulations, the 1998 Waste Discharge Regulations, Water Supply Regulations of 1999, Sewerage Regulations of 1999, National Environment Management Authority Act Cap. 153, National Water & Sewerage Corporation Statute Cap 317, Uganda Water Action Plan of 1995, National Water Policy of 1999, National Environment Management Policy of 1994, National Health Policy and Health Sector Strategic Plan of 1999, the 1997 National Gender Policy, and the Community Mobilisation and Empowerment Strategy of 2006.

Since Uganda ascribes to international water conventions and declarations, it has prepared a Water Action Plan (WAP). The WAP is a flexible and dynamic framework for the protection and development of Uganda's water resources in conformity with the Water Act.

The government of Uganda further developed a water policy that provides for the overall policy framework for water resources management and development. The overall objective of the water policy is *"To manage and develop the water resources of Uganda in an integrated and sustainable manner, so as to secure and provide water of adequate quantity and quality for all social and economic needs of the present and future generations with the full participation of all stakeholders"* [19].

The National Water Policy for Uganda identifies a need to enhance co-ordination and collaboration between the water and sanitation sub-sectors, so that they both adequately address environmental health and sanitation issues. This is done through the Water Policy Committee (WPC), Water Sector Working group (WSWG), Donor Forum, Non-Government Organisations (NGOs) forum under the umbrella of Uganda Water and Sanitation Network (UWASNET), Sector reviews by GoU and Donors and District Water and Sanitation Committee (DWSC).The policy also acknowledges the need for cooperation on trans boundary water resources management issues and promotes decentralization of water management functions. Water supply and sanitation, water for agriculture, hydropower and industrial use are priorities areas for allocation of water through planning and operation of systems.

The key player in water governance is the Ministry of Water and Environment (MWE), which has the overall responsibility of setting national policies and standards, and priorities for water development and management. It also monitors and evaluates sector development programmes to keep track of their performance, efficiency and effectiveness in service delivery. Under the Ministry there are two agencies for water sector namely; DWD as the lead agency responsible for managing water resources, coordinating and regulating all water and sanitation activities and providing support services to local Governments and other service providers. The National Water and Sewerage Corporation (NWSC) an autonomous parastatal entity established in 1972, and recognised by NWSC statute of 1995, is responsible for the delivery of water supply and sewerage services.

The other players in water governance are Ministry of Finance, Planning and Economic Development (MoFPED) mobilises funds, the Ministry of Health (MoH) is responsible for hygiene and sanitation promotion for households, while the Ministry of Education and Sports (MoES) does similar work in primary schools, the Ministry of Gender, Labour and Social Development (MoGLSD) is responsible for gender responsiveness and community mobilisation, the Ministry of Agriculture, Animal Industry and Fisheries (MAAIF)spearheads agricultural development. This includes the use and management water for production including Irrigation, animal production and aquaculture at farm level. Local Governments (Districts, towns, Sub-Counties) are empowered by the Local Governments Act of 1997 to provide water services. Communities are responsible for demanding, planning, contributing a cash contribution to operating and maintaining rural WSS facilities. While (NGOs) and Community Based Organisations(CBOs) are active in the provision of water and sanitation services, training of communities and local governments, hygiene promotion as well as advocacy and lobbying.

Since 1993, Uganda has been undergoing a decentralization of responsibilities from central government level to district local government level. This shift of responsibility has had a profound impact on the players in the Water Supply and Sanitation (WSS) sector. Districts became the main implementation agents and the human dimension is being emphasised in governance issues.

8. Catchment Based Water Resources Management

There is a lot of diversity presented in literature relating to the implementation of Integrated Water Resources Management (IWRM), especially in developing countries. This diversity is attributed to differences in institutional set up, governance structures, legal framework and level of development. The framework of IWRM stipulates that water resources are better managed at a catchment unit. This is why the WRM sub-sector reform study in Uganda recommended a paradigm shift from a centralized to a catchment based water resource management. Uganda is divided into four main management zones namely; Victoria, Kyoga, Albert and Upper Nile water management zones.

A pilot study on river Rwizi catchment area located in Victoria water management zone has revealed some aspects in the initiation of catchment management. Such as identification of threatened wetlands systems, integration of catchment management issues in district development plans, promotion of community participation, forestation, and dissemination of weather outlook/forecasts among others [40].

One of the key lessons learnt during piloting was that IWRM is better embraced in an area that is experiencing serious water resources problems. Secondly, because the process requires involvement of multiple stakeholders for its success, it involves a great deal of consultations and consequently substantial amounts of funds are required. Involving political, administrative and technical representatives in catchment management structures was found to be a viable aspect, thus it is highly encouraged.

9. Water Resources Management Challenges and Issues

Water resources in Uganda are estimated at 66 km^3/year corresponding to about 2800 m^3/person/year. The spatial and temporal distribution of water resource is uneven, which pauses a big challenge to their management. Some areas like those in north-eastern Uganda have less water resources while those in the central have plenty of water resources. That is why there are increasing incidences of water use conflicts in the water scarce parts of the country, especially in the cattle corridor. Pastoralists migrate from place to place in search of water and vegetation. This movement not only creates a security threat, it is also a major health hazard as diseases are transferred from one part of the country to another in this process.

There is also increasing pressure on water resources due to rapid population growth, increased urbanization and industrialization, uncontrolled environmental degradation and pollution. A combination of some of these factors is cited to be responsible for the recent landslides in eastern Uganda. This pressure still remains a big challenge to the sustainable management and development of country's water resources.

The frequent recurrence of extreme weather events (floods and droughts) and increasingly erratic rainfall are a big challenge to the management of Uganda's water resources. The variability in seasonal rainfall, has significantly affected the different socio-economic activities that are heavily dependent on rainfall.

Ground water is presently the major source of rural domestic water supply and is planned as the main source of water for small towns (40 completed and 80 new ones) and rural growth centres (800) by 2015. Some of the aquifers however, are limited in yield, extent, hydraulic characteristics, and recharge is low in certain parts of the country. The availability and quality of groundwater for larger rural water supply projects is a significant future challenge. Given the limitations of ground water supplies, it would appear that groundwater abstractions for many large projects such as irrigation and municipal water supplies might not be sustainable. There is still little knowledge on how climate change will impact on this important resource.

Given the increasing demand for water and the high rates of population growth of 3.2%, it is of utmost importance that an analysis of the hydrological impacts of climate change for the country are intensified.

Despite all the above planned developments, there is still very limited knowledge of the country's groundwater resources, making it difficult to guarantee sustainable groundwater development for the current and future needs.

Groundwater recharge varies considerably across the country and is extremely sensitive to land use and the amount and intensity of precipitation falling in a given area. However, due to inadequate data and resources, very few groundwater recharge assessments have been carried out in Uganda and thus recharge estimates for most areas remain unknown.

Government should be aware that, the development of hydropower schemes, though a non-consumptive use, creates reservoirs, which have adverse effects and environmental consequences on water resources. These include; Siltation/sedimentation and aquatic weed encroachment in the reservoir, diminished downstream river discharge, eutrophication in reservoirs due to domestic and industrial waste, increased loss of water due to evaporation and others.

The unique geographical location of Uganda in the Nile Basin, *i.e.* being a lower and an upper riparian state imposes constraints and responsibilities to the country [41]. As a country it would be affected by the use of the water both in quantity and in quality by the riparian countries.

Another setback in Uganda's effort especially to exploit HEP potential is the current colonial agreement of the agreed curve policy [41]. This colonial agreement has been rejected by Uganda even though the main beneficiaries of the agreement are still upholding it. It is therefore expected that regulation of Lake Victoria with a more scientific friendly policy supported by Decision Support Systems (DSS) may overcome this problem.

The challenges that Uganda faces in this respect include: failure to regulate Lake Victoria with a better regulatory policy, fluctuation of levels of lake Victoria and other equatorial lakes in the Nile Basin due to climatic change and variability, maintenance of water quality to be free of aversive weeds, sedimentation etc.

10. Conclusions

Priority of Uganda's water resources utilization and development is given to domestic water supply for both human consumption and livestock. Rural domestic water demand is by far greater than the urban demand because about 80% of the country's population is rural. Currently the rural population is dependent on ground water development or rainwater harvesting to meet their needs. Given the MDG water targets and GoU water development programmes, Uganda will resort to the utilization of the watercourses and the shared water bodies to satisfy rural water demand sooner than later. The water supply sector is under expansion and small-scale irrigation is being promoted and may in the future be of increased importance, but rainfall, which is a major source of water, is on a decrease as water demand increases. Detailed assessments are thus necessary on ground water resources, groundwater recharge and water balance.

Evidence shows that, rate of access to domestic water for rural and urban water populations had levelled off at 63% and 77% for the year 2007/2008 [24]. Major source of water for the country are the springs and wells. The new national strategy of bulk water supply is to meet rural settings demands including demands of water for livestock (WSS sector working group report, 2006). The biggest challenge is that this demand is not quantified, not forgetting that Uganda is both a downstream and an upstream riparian state.

References

[1] Kumar, R., Singh, R.D. and Sharma, K.D. (2005) Water Resources of India. *Current Science*, **89**, 794-811.

[2] Kundzewicz, Z.W., Mata, L.J., Arnell N.W., Döll, P., Kabat, P., Jiménez, B., Miller, K.A., Oki, T., Sen, Z. and Shiklomanov, I.A. (2007) Freshwater Resources and Their Management. Climate Change 2007: Impacts, Adaptation and Vulnerability. In: Parry, M.L., Canziani, O.F., Palutikof, J.P., van der Linden, P.J. and Hanson, C.E., Eds., *Contribution of Working Group* 11 *to the Fourth Assessment Report of the Intergovernmental Panel on Climate Change*, Cambridge University Press, Cambridge, 173-210.

[3] WRMD (2004) The Year-Book of Water Resources Management Department (WRMD) 2002-2003. Entebbe.

[4] UN-WWAP (2006) Uganda National Water Development Report; Prepared for the 2nd UN World Water Development Report "Water a Shared Responsibility" UN-WATER, WWAP/2006/9. World Water Assessment Programme (WWAP).

[5] Asadullah, A., Mcintyre, N. and Kigobe, M. (2008) Evaluation of Five Satellite Products for Estimation of Rainfall over Uganda. *Hydrological Sciences Journal*, **53**, 1137-1150. http://dx.doi.org/10.1623/hysj.53.6.1137

[6] NEMA (2008) State of Environment Report for Uganda. National Environment Management Authority (NEMA), Kampala.

[7] Anyah, R.O. and Semazzi, F.H.M. (2004) Simulation of the Sensitivity of Lake Victoria Basin Climate to Lake Surface Temperatures. *Theoretical and Applied Climatology*, **79**, 55-69. http://dx.doi.org/10.1007/s00704-004-0057-4

[8] FAO (2005) Uganda-AQUASTAT Survey, Irrigation in Africa in Figures. FAO Water Reports, Rome. http://www.fao.org/nr/water/aquastat/countries_regions/uganda/index.stm

[9] NEMA (1998) State of Environment Report for Uganda. National Environment Management Authority (NEMA), Kampala.

[10] DWRM (2011) A Case Study of River Rwizi, Lake Wamala, Lake Victoria Catchments and Representative Ground Water Monitoring Stations. Water Resources Monitoring and Assessment Division. Directorate of Water Resources

Management (DWRM) Entebbe, Uganda.

[11] Mölg, T., Georges, C. and Kaser, G (2003) The Contribution of Increased Incoming Shortwave Radiation to the Retreat of the Rwenzori Glaciers, East Africa, during the 20th Century. *International Journal of Climatology*, **23**, 291-303. http://dx.doi.org/10.1002/joc.877

[12] Awange, J.L., Ogalo, L., Bae, K.-H., Were, P., Omondi, P., Omute, P. and Omullo, M. (2008) Falling Lake Victoria Water Levels: Is Climate a Contributing Factor? *Climate Change*, **89**, 281-297. http://dx.doi.org/10.1007/s10584-008-9409-x

[13] Mileham, L., Taylor, R.G., Todd, M., Tindimugaya, C. and Thompson, J. (2009) The Impact of Climate Change on Groundwater Recharge and Runoff in a Humid, Equatorial Catchment: Sensitivity of Projections to Rainfall Intensity. *Hydrological Sciences Journal*, **54**, 727-738. http://dx.doi.org/10.1623/hysj.54.4.727

[14] Taylor, R.G. and Howard, K.W.F. (1996) Groundwater Recharge in the Victoria Nile Basin of East Africa: Support for the Soil-Moisture Balance Method Using Stable Isotope and Flow Modelling Studies. *Journal of Hydrology*, **180**, 31-53. http://dx.doi.org/10.1016/0022-1694(95)02899-4

[15] Tate, E., Suitcliffe, J., Conway, D. and Farquharson, F. (2004) Water Balance of Lake Victoria: Update to 2000 and Climate Change Modelling to 2100. *Hydrological Sciences Journal*, **49**, 563-574.

[16] Tindimugaya, C. (2006) Overview of Groundwater Development in Uganda. *Proceedings of the Workshop for Groundwater Professionals in Uganda*, Kampala, 25 August 2006, Unpublished.

[17] Nyenje, P.M. and Batelaan, O. (2009) Estimating the Effects of Climate Change on Groundwater Recharge and Baseflow in the Upper Ssezibwa Catchment, Uganda. *Hydrological Sciences Journal*, **54**, 713-726. http://dx.doi.org/10.1623/hysj.54.4.713

[18] Tindimugaya C (2005) Groundwater Resources Management in Urban Areas of Uganda: Experiences and Challenges. Conference Paper Maximizing the Benefits from Water and Environmental Sanitation. *Proceedings of the 31st WEDC Conference*, Kampala, 31 October-4 November 2005, 311-313.

[19] Ministry of Water, Lands and Environment (1999) The National Water Policy. MWLE, Kampala.

[20] NBCBN (2005) Small Scale Hydropower for Rural Development. Hydropower Development Research Cluster, Group 1, Nile Basin Capacity Building Network NBCBN-RE.

[21] ERA (2009) Development and Investment Opportunities in Renewable Energy Resources in Uganda. Electricity Regulatory Authority, Kampala.

[22] Umeme (2011) Umeme Annual Report and Financial Statements for the Year Ended December 2011. Kampala.

[23] Kasita, I. (2011) Electricity Demand to Triple by 2023. The New Vision, Kampala.

[24] MOFPED (2011) Annual Economic Performance Report 2010/11. Directorate of Economic Affairs, Ministry of Finance, Planning and Economic Development, Kampala.

[25] UBOS (2009) Uganda Bureau of Statistics, 2009 Statistical Abstracts. Kampala. www.ubos.org.

[26] African Development Bank (2010) The African Development Bank in Action, Activities in the Water and Sanitation Sector in Uganda: Overview and Key Elements of Interventions. Water and Sanitation Department of the African Development Bank. http://www.afdb.org/fileadmin/uploads/afdb/documents/Projects-and-Operations/9_AFDB_watsan_UG

[27] African Development Bank (2012) AFDB Promotes Improved Access to Water, Sanitation and Health Services in Rural and Urban Uganda. African Development Bank Group. http://www.afdb.org/en/news-and-events/article/afdb-promotes-improved-access-to-water-sanitation-and-health-services-in-rural-and-urban-uganda-8771/

[28] Kull, D. (2006) Connections between Recent Water Level Drops in Lake Victoria, Dam Operations and Drought. http://www.internationalrivers.org/files/attached-files/full_report_pdf.pdf

[29] Hulme, M. (1992) Rainfall Changes in Africa: 1931-1960 to 1961-1990. *International Journal of Climatology*, **12**, 685-699. http://dx.doi.org/10.1623/hysj.54.4.713

[30] Hastenrath, S. (2001) Variations of East African Climate during the Past Two Centuries. *Climatic Change*, **50**, 209-217. http://dx.doi.org/10.1023/A:1010678111442

[31] Mistry, V.V. and Conway, D. (2003) Remote Forcing of East African Rainfall and Relationships with Fluctuations in Levels of Lake Victoria. *International Journal of Climatology*, **23**, 67-89. http://dx.doi.org/10.1002/joc.861

[32] Nicholson, S.E., Yin, X. and Ba, M.B. (2000) On the Feasibility of Using a Lake Water Balance Model to Infer Rainfall: An Example from Lake Victoria. *Hydrological Science Journal*, **45**, 75-95. http://dx.doi.org/10.1080/02626660009492307

[33] Taylor, R.G., Mileham, L., Tindimugaya, C. and Mwebembezi, L. (2009) Recent Glacial Recession and Its Impact on

Alpine River Flow in the Rwenzori Mountains of Uganda. *Journal of African Earth Sciences*, **55**, 205-213.
http://dx.doi.org/10.1016/j.jafrearsci.2009.04.008

[34] Kaser, G., Hardy, D.R., Mölg, T., Bradley, R.S. and Hyera, T.M. (2004) Modern Glacier Retreat on Kilimanjaro as Evidence of Climate Change: Observations and Facts. *International Journal of Climatology*, **24**, 329-339.
http://dx.doi.org/10.1002/joc.1008

[35] MWE (2011) Uganda Water and Environment Sector Performance Report 2011, Ministry of Water, Lands and Environment, Kampala.

[36] GWP (2000) Integrated Water Resources Management. Global Water Partnership (GWP) Technical Advisory Committee, Background Paper No.4.

[37] Rogers, P. and Hall, A.W. (2003) Effective Water Governance. Technical Committee Background Papers No.7, Global Water Partnership (GWP).

[38] Turton, A.R., Hattingh, H.J., Maree, G.A., Roux, D.J., Claassen, M. and Strydom, W.F., Eds. (2007) Governance as a Trialogue: Government-Society-Science in Transition. Springer-Verlag Berlin, Heidelberg.
http://dx.doi.org/10.1007/978-3-540-46266-8

[39] Sewagudde, S. (2011) Catchment Based Water Resources Management: Uganda's Journey to Taking the IWRM Concept to the Grass Root. *Wash Watch*, **2**, 19-23.

[40] Human Development Report, HDR (2006) Beyond Scarcity: Power, Poverty and Global Water Crisis. Palgrave Macmillan, New York.

[41] NEMA (1996) State of the Environment Report for Uganda 1996. National Environment Management Authority (NEMA), Kampala.

Permissions

All chapters in this book were first published in JWARP, by Scientific Research Publishing; hereby published with permission under the Creative Commons Attribution License or equivalent. Every chapter published in this book has been scrutinized by our experts. Their significance has been extensively debated. The topics covered herein carry significant findings which will fuel the growth of the discipline. They may even be implemented as practical applications or may be referred to as a beginning point for another development.

The contributors of this book come from diverse backgrounds, making this book a truly international effort. This book will bring forth new frontiers with its revolutionizing research information and detailed analysis of the nascent developments around the world.

We would like to thank all the contributing authors for lending their expertise to make the book truly unique. They have played a crucial role in the development of this book. Without their invaluable contributions this book wouldn't have been possible. They have made vital efforts to compile up to date information on the varied aspects of this subject to make this book a valuable addition to the collection of many professionals and students.

This book was conceptualized with the vision of imparting up-to-date information and advanced data in this field. To ensure the same, a matchless editorial board was set up. Every individual on the board went through rigorous rounds of assessment to prove their worth. After which they invested a large part of their time researching and compiling the most relevant data for our readers.

The editorial board has been involved in producing this book since its inception. They have spent rigorous hours researching and exploring the diverse topics which have resulted in the successful publishing of this book. They have passed on their knowledge of decades through this book. To expedite this challenging task, the publisher supported the team at every step. A small team of assistant editors was also appointed to further simplify the editing procedure and attain best results for the readers.

Apart from the editorial board, the designing team has also invested a significant amount of their time in understanding the subject and creating the most relevant covers. They scrutinized every image to scout for the most suitable representation of the subject and create an appropriate cover for the book.

The publishing team has been an ardent support to the editorial, designing and production team. Their endless efforts to recruit the best for this project, has resulted in the accomplishment of this book. They are a veteran in the field of academics and their pool of knowledge is as vast as their experience in printing. Their expertise and guidance has proved useful at every step. Their uncompromising quality standards have made this book an exceptional effort. Their encouragement from time to time has been an inspiration for everyone.

The publisher and the editorial board hope that this book will prove to be a valuable piece of knowledge for researchers, students, practitioners and scholars across the globe.

List of Contributors

Haiyan Peng and Niels Thevs
Institute of Botany and Landscape Ecology, University of Greifswald, Greifswald, Germany

Konrad Ott
Department of Philosophy, Christian-Albrecht-University Kiel, Kiel, Germany

Anupam Khajuria, Sayaka Yoshikawa and Shinjiro Kanae
Department of Civil Engineering, Tokyo Institute of Technology, Tokyo, Japan

Maisa'a W. Shammout
Water, Energy and Environment Center, The University of Jordan, Amman, Jordan

Douglas N. Kastendick, Brian J. Palik, Randy K. Kolka and Joshua J. Kragthorpe
USDA Forest Service, Northern Research Station, Grand Rapids, USA

Eric K. Zenner
Penn State University, University Park, USA

Charles R. Blinn
University of Minnesota, St. Paul, USA

Li Fu
Shanghai Guanglian Construction Development Co. Ltd., Shanghai, China

Javier Osorio, Jaehak Jeong and Katrin Bieger
Blackland Research and Extension Center, Texas A & M AgriLife Research, Temple, USA

Jeff Arnold
Grassland, Soil and Water Research Laboratory, United States Department of Agriculture, Agricultural Research Service (USDA-ARS), Temple, USA

Nara Somaratne and Glyn Ashman
South Australian Water Corporation, Adelaide, Australia

Jeff Lawson
Department of Environment, Water and Natural Resources, Mount Gambier, Australia

Kien Nguyen
Hydraulic Works and Management Division, Directorate of Water Resources, Ministry of Agriculture and Rural Development, Hanoi, Vietnam

Mohammad Mirzavand, Hoda Ghasemieh, Rasool Imani and Mehdi Soleymani Motlagh
Department of Natural Resources and Geoscience, University of Kashan, Kashan, Iran

Seyed Javad Sadatinejad
Department of New Sciences and Technologies, University of Tehran, Tehran, Iran

Nadhir Al-Ansari, Ammar A. Ali and Sven Knutsson
Department of Civil, Environmental and Natural Resources Engineering, Lulea University of Technology, Lulea, Sweden

Luan Pan, Viacheslav I. Adamchuk and Shiv O. Prasher
Department of Bioresource Engineering, McGill University, Ste-Anne-de-Bellevue, Canada

Richard B. Ferguson
Department of Agronomy and Horticulture, University of Nebraska-Lincoln, Lincoln, USA

Pierre R. L. Dutilleul
Department of Plant Science, McGill University, Ste-Anne-de-Bellevue, Canada

Hector German Rodriguez and Jennie Popp
Department of Agricultural Economics and Agribusiness, University of Arkansas, Fayetteville, USA

Edward Gbur
Agricultural Statistics Laboratory, University of Arkansas, Fayetteville, USA

John Pennington
Beaver Watershed Alliance, Fayetteville, USA

M. Edwin S. Lubis, I. Yani Harahap, Taufiq C. Hidayat, Y. Pangaribuan and Edy S. Sutarta
Indonesia Oil Palm Research Institute, Medan, Indonesia

Zaharah A. Rahman, Christopher The and M. M. Hanafi
Universiti Putra Malaysia, Serdang, Malaysia

Awatif S. Alsaqqar, Basim H. Khudair and Sura Kareem Ali
Department of Civil Engineering, College of Engineering, Baghdad University, Baghdad, Iraq

Francis N. W. Nsubuga
Department of Geography, Geoinformatics and Meteorology, University of Pretoria, Pretoria, South Africa
National Environmental Consult Ltd., Kampala, Uganda

Masoud Nsubuga-Ssenfuma
National Environmental Consult Ltd., Kampala, Uganda

Edith N. Namutebi
Ministry of Foreign Affairs, Kampala, Uganda